Civil Liability for Damage Caused by Global Navigation Satellite System

Aerospace Law and Policy Series

VOLUME 15

Editors

Pablo Mendes de Leon, Professor of Air and Space Law, and Tanja Masson-Zwaan, Assistant Professor of Space Law, Leiden University, The Netherlands.

Introduction

The *Aerospace Law and Policy Series* critically examines the fundamental changes that international aviation and space activities have undergone since the last century, growing up as mature industries while freeing themselves from traditional regulatory constraints and displaying a variety of innovative applications that call for creative legal solutions.

Objective

The objective of the Series is to make a contribution to legal thinking on public international air and space law and policy, and their implementation at the regional and national levels, responsibility and liability of public bodies and service providers, competition law, insurance law, company law, and the complex relationship between EU law and public international law.

Readership

Aviation and space lawyers, academics, representatives of governments, international organizations, and companies involved in aviation and space activities.

The titles published in this series are listed at the end of this volume.

Civil Liability for Damage Caused by Global Navigation Satellite System

Dejian Kong

Wolters Kluwer

Published by:
Kluwer Law International B.V.
PO Box 316
2400 AH Alphen aan den Rijn
The Netherlands
E-mail: international-sales@wolterskluwer.com
Website: lrus.wolterskluwer.com

Sold and distributed in North, Central and South America by:
Wolters Kluwer Legal & Regulatory U.S.
7201 McKinney Circle
Frederick, MD 21704
United States of America
Email: customer.service@wolterskluwer.com

Sold and distributed in all other countries by:
Air Business Subscriptions
Rockwood House
Haywards Heath
West Sussex
RH16 3DH
United Kingdom
Email: international-customerservice@wolterskluwer.com

ISBN 978-94-035-1171-9

e-Book: ISBN 978-94-035-1233-4
web-PDF: ISBN 978-94-035-1234-1

To my daughter Li Kong (孔礼)

Supported by Program for Young Innovative Research Team
in China University of Political Science and Law

About the Author

Dr Dejian Kong is an assistant professor of law at the China University of Political Science and Law, and a fellow of the Research Institute of Air and Space Law. He was born in Shandong Province, China. He received his Bachelor's degree in law from Shandong University, China and passed his bar examination in 2010. During his undergraduate study, he was awarded National Scholarship twice and won the honorary title of 'Outstanding Graduates of Shandong Province'. At the same year, he was admitted, with an exemption from entrance examination, as a master student with a full scholarship by School of Law, Beihang University (formerly known as Beijing University of Aeronautics and Astronautics), China. His master thesis entitled 'Legal Study on the Possessory Lien over Civil Aircraft' was awarded as the best LLM thesis in 2013 when he got a master degree. He then was selected by the China Scholarship Council – Leiden University Joint Scholarship Programme and started his PhD research at the International Institute of Air and Space Law, Leiden University, The Netherlands with the supervision from Professor Pablo Mendes de Leon (Leiden University) and Professor Anna Masutti (the Unversity of Bologna, Italy).

During his stay in Leiden, Mr Kong had been focused on the law of Global Navigation Satellite System (GNSS), particularly the issue of civil liability in the context of GNSS. He visited the University of Bologna (Italy), the European Institute of Space Policy (Austria), and the International Institute for the Unification of Private Law (Italy) for academic purpose. Meanwhile, he served as a research assistant at the National Research Center of Air Traffic Management Law and Standard of China, the Institute of Air Law and Standard of Beihang University. He is also active in academic events and publications. One of his papers was awarded as the Best-in-Grade Award of 2018 by China Institute of Space Law.

On 17 December 2018, Mr Kong successfully presented and defended his PhD thesis and was awarded with a doctoral degree by Leiden University.

Table of Contents

List of Abbreviations

AAI	Airports Authority of India
ABAS	Aircraft-Based Augmentation System
ADS-B	Automatic Dependent Surveillance-Broadcast
Aerospace	Air and Space
ANS	Air Navigation Service
ANSP	Air Navigation Services Provider
ASAS	African Satellite Augmentation System
ATC	Air Traffic Control
ATM	Air Traffic Management
ATS	Air Traffic Service
BDS	BeiDou Navigation Satellite System
CANSO	Civil Air Navigation Services Organisation
CGSIC	Civil GPS Service Interface Committee
Chicago Convention	Convention on International Civil Aviation
CNS/ATM	Communications, Navigation and Surveillance/Air Traffic Management
COPUOS	Committee on the Peaceful Uses of Outer Space
CORS	Continuously Operating Reference Station
CSNO	China Satellite Navigation Office
DHS	US Department of Homeland Security
EC	European Commission
ECAC	European Civil Aviation Conference
EDAS	EGNOS Data Access Service
EGNOS	European Geostationary Navigation Overlay Service
eLoran	Enhanced Loran
ESA	European Space Agency

ESDP	European Security and Defence Policy
ESPI	European Space Policy Institute
ESSP SAS	European Satellite Services Provider
EU	European Union
EUROCONTROL	European Organisation for the Safety of Air Navigation
EUTELSAT IGO/Eutelsat	European Telecommunications Satellite Organization
EWA	EGNOS Working Agreements
FAA	US Federal Aviation Administration
FANS	Future Air Navigation System(s)
FANS Phase II	Special Committee for the Monitoring and Coordination of Development and Transition Planning for the Future Air Navigation System
FCC	US Federal Communications Commission
FIR	Flight Information Regions
GAGAN	GPS-aided GEO Augmented Navigation
Galileo	Galileo Satellite Navigation System
GANP	Global Air Navigation Plan
GAO	US Government Accountability Office
GBAS	Ground-Based Augmentation System
GLONASS	Globalnaya Navigatsionnaya Sputnikovaya Sistema
GLS	GNSS Landing System
GNSS	Global Navigation Satellite System
GNSSP	Global Navigation Satellite System Panel
GPG	Global/World Public Good
GPS	Global Positioning System
GSA	European GNSS Agency
IATA	International Air Transportation Association
ICAO	International Civil Aviation Organization
ICG	International Committee on Global Navigation Satellite Systems
ICJ	International Court of Justice
ICRC	International Committee of the Red Cross
ICT	Information and Communication Technology
IFATCA	International Federation of Air Traffic Controllers' Associations
iGMAS	International GNSS Monitoring and Assessment
IGO	Inter-Governmental Organization
IGS	International GNSS Service
IISL	International Institute of Space Law
ILA	International Law Association
ILC	International Law Commission

IMO	International Maritime Organization
IMSO	International Mobile Satellite Organization
INMARSAT	International Maritime Satellite Organization
INS	Inertial Navigation System
ITSO/Intelsat	International Telecommunications Satellite Organization
ION	Institute of Navigation
IRNSS	Indian Regional Navigation Satellite System
ISO	International Organization for Standardization
IS-QZSS	Interface Specification for QZSS
ISRO	Indian Space Research Organization
ISS	International Space Station
ITU	International Telecommunication Union
ITU-R	ITU Radiocommunication Sector
LAAS	Local Area Augmentation System
Liability Convention	Convention on International Liability for Damage Caused by Space Objects
LTEP	ICAO Panel of Legal and Technical Experts on the Establishment of a Legal Framework with Regard to GNSS
MLS	Microwave Landing System
Moon Agreement	Agreement Governing the Activities of States on the Moon and Other Celestial Bodies
MOU	Memorandum of Understanding
MPA	Maritime and Port Authority of Singapore
MSAS	Multi-functional Satellite Augmentation System
MSC	Maritime Safety Committee
NAVCEN	US Coast Guard Navigation Center
NCL-CBU	Non-contractual Liability vis-à-vis Contractually Bound Users, Such as Civil Liability Claimed by Commercial Services
NCL-UNC	Non-contractual Liability vis-à-vis Users Not Bound by Contract
NDGPS	Nationwide Differential GPS
NNSS	National Navigation Satellite System
NOAA	US National Oceanic and Atmospheric Administration
NOC for Space-Based PNT	National Coordination Office for Space-Based Positioning, Navigation, and Timing
NSP	Navigation System Panel
OECD	Organisation for Economic Co-operation and Development

Outer Space Treaty	Treaty on Principles Governing the Activities of States in the Exploration and Use of Outer Space, including the Moon and Other Celestial Bodies
PANS	Procedures for Air Navigation Services
PANS-ATM	Procedures for Air Navigation Services – Air Traffic Management
PANS-OPS	Procedures for Air Navigation Services – Aircraft Operations
PBN	Performance-Based Navigation
PNT	Positioning, Navigation and Timing
PPP	Public-Private Partnership
QSS	Quasi-Zenith Satellite System Services Inc.
QZSS	Quasi-Zenith Satellite System
Registration Convention	Convention on Registration of Objects Launched into Outer Space
Rescue Agreement	Agreement on the Rescue of Astronauts, the Return of Astronauts and Return of Objects Launched into Outer Space
RNAV	Area Navigation
RNSS	Regional Navigation Satellite System
Rome Convention	Convention on Damage Caused by Foreign Aircraft to Third Parties on the Surface of 1952
Rome II Regulation	Regulation (EC) No 864/2007 of the European Parliament and of the Council of 11 July 2007 on the law applicable to non-contractual obligations
Roscosmos	Russian Federal Space Agency State Corporation for Space Activities
SA	Selective Availability
SAR	Search and Rescue
SARPs	Standards and Recommended Practices
SBAS	Satellite-Based Augmentation System
SDCM	System for Differential Corrections and Monitoring
SDRs	Special Drawing Rights
SES	Single European Sky
SIS	Signal-in-Space
SMS	Safety Management Systems
SoL	Safety of Life
SPS	Standard Positioning Service
TRANSIT	Navy Navigation Satellite System
UN	United Nations
UNCC	United Nations Compensation Commission
UNGA	General Assembly of the United Nations

UNIDROIT	International Institute for the Unification of Private Law
UNISPACE III	Third United Nations Conference on the Exploration and Peaceful Uses of Outer Space
UNOOSA	United Nations Office for Outer Space Affairs
US	United States
Vienna Convention	Vienna Convention on the Law of Treaties
WAAS	Wide Area Augmentation System
Warsaw Convention	Convention for the Unification of Certain Rules Relating to International Carriage by Air of 1929

Acknowledgements

This book is based on the author's doctoral thesis which was done at the International Institute of Air and Space Law, Leiden University, The Netherlands, with the financial support from the China Scholarship Council – Leiden University Joint Scholarship Programme and the National Research Center of Air Traffic Management Law and Standard of China. This book could not have been done without continuous help and trust from the government, universities, friends, and family.

The deepest gratitude goes to the author's supervisors, Professor Pablo Mendes de Leon and Professor Anna Masutti for their professional and dedicated cultivation with patience and encouragement during the whole process of the author's PhD research. The author's sincere thanks go further to Drs Tanja Masson-Zwaan for her academic advisement. A special thanks to Ms Paula van der Wulp and Ms Natascha Meewisse, for all administrative assistance to make the author's research and life much easier in Leiden.

The author appreciates all the inputs to the arrangement of his PhD defence. Special thanks to Professor Clementine Breedveld-de Voogd, Professor Wolf Müller-Rostin, Dr Bart Custers, and Professor Jiefang Huang, who are members of the author's doctorate committee; to Ms Esmee ten Donkelaar for her help on all administrative matters; to Ms Neta Palkovitz Menashy for providing thoughtful comments; to Mr Stephen Machon and Mr Roberto Cassar for editorial assistance in the English language; to Ms Magdalena (Maggie) Kucko for her translation of summary into Dutch in the original thesis submitted for defence.

The author would like to express his thanks to all the academic organisations for hosting him as a full-time or part-time researcher with a longer or shorter term. Special graduate is owed to Leiden University (The Netherlands), Beihang University (China), the University of Bologna (Italy), the European Institute of Space Policy (Austria), the International Institute for the Unification of Private Law (Italy), and China Academy of Strategy on Aerospace Engineering Science and Technology, for all the opportunities offered to him.

It would not be possible for the author to study outside China without the inspiration and encouragement from Professor Baoyu Liu and Mrs Yanyan Li, Dr Hao

Liu and Dr Caixia Yang. The author's further thanks go to Dr Weihua Zhang, Ms Chen Chen, Ms Huai Yu, Ms Yun Deng, Ms Xin Zhang, Ms Bingjie Liang, and Dr Fabio Tronchetti, for their support and advice from Beijing.

The author is lucky to meet many friends for offering him a positive atmosphere. Particularly, the author wishes to thank Dr Lalin Kovudhikulrungsri, Ms Jinyoung Choi, Dr Jianxin Yang, Ms Xiaoqian Zhang, Dr Qiuyin Hu, Dr Boyang Liu, Dr Xiang Li, Dr Yudan Tan, Dr Linlin Sun, Ms Wanlu Zhang, Dr Juan Wang, Mr Julian Rotter, and Ms Marilena Montanari. The author also remember ideas inspired by young Chinese generations during lunch meetings in Kamerlingh Onnes Building. Although the author may not be able to list all the names of his friends, the author appreciates the friendship with them and all the help from them.

Only being with the family, a PhD journey may become a true life. The author's greatest love goes to his wife Mrs Zhenjie Li (李 振 杰) not only for her always being with him and limitless understanding, but also for the happiness brought by the birth and growing up of their lovely daughter Li Kong (孔 礼) since January 2017. The author wishes to express the sincerest gratitude to his parents and his parents-in-law. The author also remember the support from his sister and sister-in-law and their families.

General Introduction

Benefits generated by Global Navigation Satellite System (GNSS) have penetrated to every corner of the earth. National security, economic growth, and transportation safety as well as efficiency are severely dependent on positioning information, navigation capabilities, and time dissemination provided by GNSS. Nevertheless, that is only one side of the coin. Without exception, GNSS is not risk-free, even though, often, new technologies present opportunities for increased safety, security, and efficiency. The more humanity depends on GNSS; the more risks it has to face. A defect in or loss of GNSS signals occurring due to unintentional or intentional causes may not only endanger hundreds of lives on-board an aircraft which is based on a GNSS landing system (GLS), but it may also cause substantial economic loss to critical civilian infrastructure for banking, electricity, and other purposes.

Who is it that shall be responsible or liable in the event that real damage is unfortunately caused by GNSS to its users or any third parties, and how shall such be responsible or liable? In the case where GNSS signals are provided free of charge, is it fair to force a GNSS provider to bear the burden of compensation? Could a GNSS provider release its civil liability by the doctrine of State Immunity if that provider is a public authority? The international community has contributed decades of efforts into searching for answers for those questions, yet thus far, no consensus has been reached. Some believe that there are no matters that exist in a legal vacuum, and that the current international law regime could deal with the issue of GNSS civil liability; others confirm serious doubts on the fitness of 'old' laws to such 'new' technologies, the reasoning being that, when stakeholders were fighting intensely for the adoption of relevant conventions, they may not have even anticipated the essential role and, in particular, the risks of GNSS.

Against this backdrop, the aim of this book is to explore whether current international law is adequate to deal with the issue of civil liability in the context of GNSS. In other words, the aim is to explore whether present international law can ensure that parties suffering damage get fair, prompt, and adequate compensation while balancing the interests of the GNSS industry in order for it to maintain its sustainable development. If so, how does international law apply in a case related to GNSS civil liability? If it does not apply, where is the legal gap and how should we move forward? For those answers, the author examines published papers and recent developments of authors, analyses, travaux préparatoires, and current texts of treaties as well as national legislation, and addresses conference documents released by international organisations. To make the analysis easily understandable, the author designed a hypothetical case where damage is caused by GNSS due to different reasons, and illustrated each legal relationship and causal link in various figures.

The book divides its contents into five chapters. All chapters start from an introduction and close with concise concluding remarks. Chapter 1 shifts its focus from GNSS technical matters to legal issues in general, and then gradually to civil liability in particular. Chapter 2 concerns a conceptual analysis of GNSS civil liability, serving as a basis for any further study. Chapters 3 and 4 check the application and adequacy of current international air and space laws to the case where damage is caused by GNSS. Chapter 5 tries to find a roadmap for the issue of GNSS civil liability to move forward in a feasible way.

Chapter 1 is an overview of GNSS from technical, financial, institutional, and legal perspectives. It begins with technical know-how of GNSS to understand what GNSS is and how GNSS benefits humanity. GNSS performance parameters are described to clarify the standard on examining the defection of GNSS signals. The global nature of GNSS is presented to serve as the technical basis for the need for the international law to deal with GNSS civil liability. Risks faced by the international GNSS community are analysed to lay the foundation of triggers of GNSS civil liability. Against those risks, the chapter proposes, among others, a uniform governance structure composed of an institutional framework and a legal system for GNSS.

Chapter 2 defines the term of civil liability in the context of GNSS from a more legal perspective. It examines which parts of the value chain of GNSS could be qualified as the origin of damage in this book. The terms 'liability' and 'responsibility' were distinguished with specific regard to GNSS. The concept of 'GNSS liability' is discussed separately based on criminal, administrative, and civil laws. The structure of 'GNSS civil liability' is further divided into three pillars: GNSS contractual liability, GNSS general tort liability, and GNSS product liability. Four elements, that are the parties, unreasonable acts, damage, and their causality, are analysed to see how GNSS civil liability is established. Considering the aforementioned global nature of GNSS from a technical perspective, the chapter continues to discuss how that factor influences the concept of GNSS civil liability in a legal sense.

Chapter 3 checks whether and how international space law applies to the issue of GNSS civil liability. It establishes a link between GNSS and international space law by reasoning that GNSS is a space system, and relevant activities belong to space activities. After presenting various sources of international space law, the chapter examines the possibility of applying both Article VII of the Treaty on Principles Governing the Activities of States in the Exploration and Use of Outer Space, including the Moon and Other Celestial Bodies (Outer Space Treaty), and the Convention on International Liability for Damage Caused by Space Objects (Liability Convention) to GNSS civil liability by answering the following questions: (i) does a GNSS signal qualify as a 'space object'? (ii) is damage caused by GNSS covered by outer space treaties as an 'indirect damage'? and (iii) where is the causal link, if at all, between damage and

GNSS? By producing answers to those questions, the chapter proceeds to assess practical challenges in claims for compensation under international space law for damage caused by GNSS.

Chapter 4 checks whether and how international air law applies to the issue of GNSS civil liability. The chapter establishes a link between GNSS and international air law by reasoning that GNSS is one of the critical infrastructures during the operation of an aircraft. The chapter divides its contents into three parts to discuss the application of legal documents of international air law to the issue of GNSS civil liability. First, the legal effect of regulatory materials and guidance documents generated on the platform of the International Civil Aviation Organization (ICAO) is examined to see whether they can be of any help to deal with the issue of GNSS civil liability. Second, the possibility to resolve the issue of GNSS civil liability under the international legal framework of Air Traffic Control (ATC) civil liability is checked. Third, the applicability of international treaties, concluded from Warsaw to Montreal via Rome, regarding air carriers' civil liability for damage caused by GNSS malfunction or defective Positioning, Navigation and Timing (PNT) signals through the operation of aircraft is reviewed.

Chapter 5 intends to present a highly practical solution for the issue of GNSS civil liability through a series of proposals. A fairness test on GNSS civil liability is scrutinised in conjunction with a free-charge policy for GNSS open signals. The question of whether a disclaimer of civil liability justifies GNSS providers not assuming civil liability is also addressed. The chapter further strives for a clear roadmap to make solutions for the issue of GNSS civil liability feasible: first, a legal and an institutional solution is presented against the doctrine of sovereignty immunity; second, various solutions for GNSS civil liability are assessed based on their respective feasibility; and third, several international organisations are listed to see how shall they work together to achieve the way forward for the issue of GNSS civil liability.

The final point of this book is to promote public safety by restraining the negligent activities and omissions of GNSS signal providers, and urging them to improve the stability of the satellite navigation systems, and these by establishing a clear regime of GNSS civil liability. The author acknowledges that, although for the time being a catastrophic crash due to a malfunction of GNSS is quite unlikely, this cannot be ruled out in the future when, for example, GNSS safety-of-life signals constitute essential elements for each autonomous aircraft, ship, and car. It is irresponsible to wait for an accident to happen merely to justify the need for an appropriate regime of GNSS civil liability. This book, accordingly, deserves certain significance and attention.

CHAPTER 1

Overview of GNSS: Technical Background and Legal Framework*

§1.01 INTRODUCTION

Technical issues are always a challenging area for legal researchers, but legal analysis must be based on sufficient technical understanding of a research subject. As a Chinese proverb indicates, 'Precise knowledge of self and precise knowledge of the threat leads to victory'. Therefore, first and foremost, this chapter provides initial technical know-how about GNSS by raising the question: 'what is GNSS, and how can humanity benefit from GNSS?' (*see* §1.02). In facing a complicated system which has penetrated every aspect of modern life, we however need to understand the risks that are from technical, financial, institutional, and legal perspectives, as well as the potential solutions thereto (*see* §1.03). As the purpose of legal research is to solve the problem by legal methods, this chapter emphasises the question: 'what should the legal community do to address and mitigate GNSS risks?' (*see* §1.04).

The aim of this chapter is to establish a general framework of GNSS from both a technical and legal perspective. This chapter shifts its points of focus twice: first from GNSS technical issues to GNSS legal issues in general, then from GNSS legal issues in general to civil liability in particular; the reason for this is to establish a research basis for the following chapters.

* Part of this chapter was published as: Dejian Kong, *Shaping a Uniform Governance Structure over Global Navigation Satellite System (GNSS): The Way of Risk Management*, 21(2-3) Uniform Law Review 2016, at 393-415.

§1.02 BASIC INFORMATION ABOUT GNSS

[A] PNT and GNSS

Across the globe, and not simply in GNSS power States, the PNT service has been integrated into both military combat operations and national civilian critical infrastructure for transportation, banking, electricity, and other purposes. In today's age of information, national security, economic growth, and transportation safety are largely dependent on positioning information, navigation capabilities, and time dissemination. Although now it is not the sole means of achieving PNT functions, GNSS is more accurate with wider coverage compared with its next-best ground-based alternatives, such as Enhanced Loran (eLoran).[1] Therefore, ever since the first GNSS – TRANSIT (Navy Navigation Satellite System) – was declared operational,[2] with a game-changing effect, GNSS has been the main solution for providing precise PNT data with global coverage.

[B] The Constitution of GNSS

GNSS is the generic term for all satellite navigation systems with global coverage rather than a specific system under daily operation in the real world. For both national security and economic interests, the United States (US), China, Russia, and the European Union (EU) are in various stages of deploying their own GNSSs, that is, the Global Positioning System (GPS), BeiDou Navigation Satellite System (BDS), Globalnaya navigatsionnaya sputnikovaya Sistema (GLONASS), and Galileo Satellite Navigation System (Galileo) respectively. Although only GPS and GLONASS are now fully operational, the EU and China are moving quickly towards deploying their own global systems, and the spring of GNSS is expected to arrive around the decisive year of 2020, when (a) both BDS and Galileo will become operational;[3] and (b) both GPS and GLONASS will fully broaden their new civil signals.[4] In addition, most regional navigation satellite systems (RNSSs), such as the Indian Regional Navigation Satellite System (IRNSS), have the potential of reaching global coverage by launching more satellites. Meanwhile, GNSS can extend its global coverage and international character,

1. For more information on eLoran and similar systems, *see A BILL: To Require the Secretary of Defense to Establish a Backup for the Global Positioning System, and for Other Purposes*, H.R.1678, 114th Congress (2015-2016); Gelli Manoj Someswar, et al., *Global Navigation Satellite Systems and Their Applications*, 17(1) International Journal of Software and Web Sciences 2013, at 19.
2. Norman Bonnor, *A Brief History of Global Navigation Satellite Systems*, 65 Journal of Navigation 2012, at 3.
3. *See* China Satellite Navigation Office (CSNO), *Report on the Development of BeiDou Navigation Satellite System (Version 2.2)*, December 2013; Article 3 of Regulation (EU) No 1285/2013.
4. *See* the US Department of Defense, *Preservation of Continuity for Semi-Codeless GPS Applications*, Federal Register / Vol. 73, No. 185 / Tuesday, 23 September 2008 / Notices, at 54792-54793; Paul D. Groves, *Principles of GNSS, Inertial, and Multisensor Integrated Navigation Systems* (Artech House, 2013), at 323; Kevin Roebuck, *GLONASS: High-Impact Strategies – What You Need to Know: Definitions, Adoptions, Impact, Benefits, Maturity, Vendors* (Emereo Pty Limited, 2011), at 6.

and increase its performance through augmentation systems such as the European Geostationary Navigation Overlay Service (EGNOS).[5] If a GNSS signal is augmented, the original signal is named the 'core signal' or 'basic signal', while the signal processed by the augmentation system is termed the 'augmented signal'. Analogously, a GNSS-produced core/basic signal is a 'core/basic GNSS' (*see* Figure 1.1).

Figure 1.1 GNSS Elements

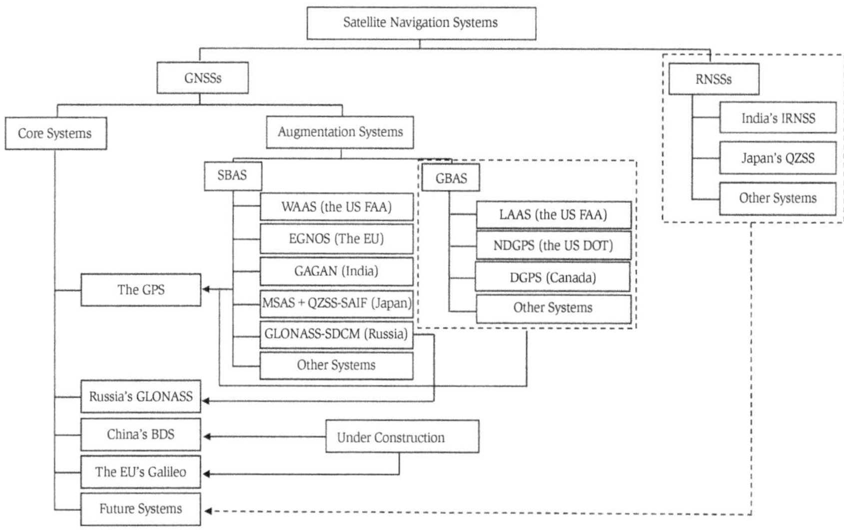

As to the technical constitution, a Satellite-Based Augmentation System (SBAS) has its own satellites; a Ground-Based Augmentation System (GBAS) works with its own ground facilities; each core GNSS is composed of the following three segments from outer space to the ground:

(i) *Space Segment*, a constellation of dozens of satellites that continually transmit radio signals containing navigation data in designated frequencies.
(ii) *Control Segment*, a ground-based network composed of master control stations, data uploading stations, and monitoring stations all over the globe.

5. At present, GNSS augmentation systems are composed of GBAS and SBAS. Most augmentation systems only enhance GPS signals, including the US WAAS, LAAS, and NDGPS, the EU's EGNOS, Japan's MSAS, India's GAGAN, Africa's ASAS, etc. However, Russia's SDCM could perform integrity monitoring of both GPS and GLONASS satellites. In addition, Japan's QZSS, as one of the RNSSs, could also provide augmentation of GPS signals by frequency L1-SAIF.
See respectively: UNOOSA, *Global Navigation Satellite Systems: Education Curriculum*, 2012, ST/SPACE/59, at iii; *About SDCM*, http://www.sdcm.ru/smglo/staticpages?version=eng &site=extern&title=about, last accessed 5 June 2015; Japan Aerospace Exploration Agency, *Quasi-Zenith Satellite System Navigation Service: Interface Specification for QZSS (IS-QZSS)*, IS-QZSS V1.6, 28 November 2014.

(iii) *User Segment*, the equipment that shows PNT information by computing GNSS signals received from at least three available navigation satellites, based on the trilateration method (*see* Figure 1.2).[6]

Figure 1.2 GNSS Constitution

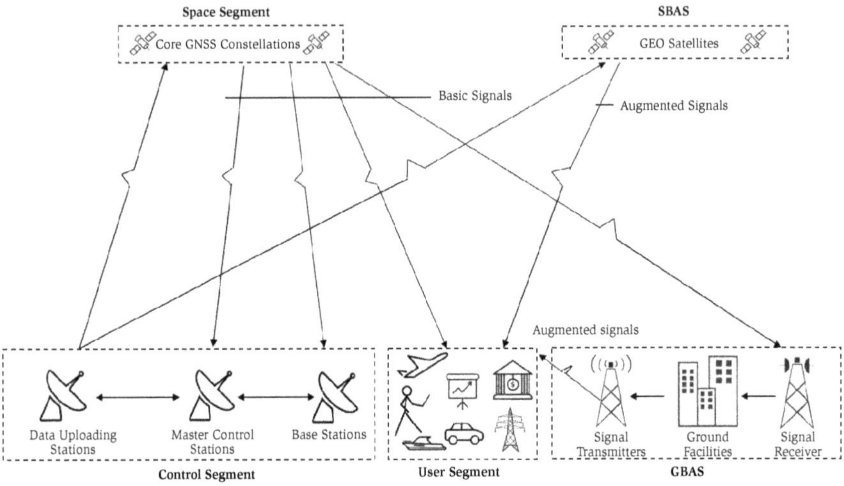

[C] Applications of GNSS

The best way to understand the role of GNSS is to imagine what would happen if GNSSs were suddenly turned off.[7] Although it would not cause panic on the streets, such an event would jeopardise all walks of life throughout the world: from vehicles on the ground, to aircraft in the air, and spacecraft or satellites in outer space; from critical infrastructures such as the grid system on the mainland, to submarines in the sea; and from software like the World Wide Web to hardware including heavy equipment for construction and mining. Yet that is not all, particularly considering that the livelihood of millions of people involved in the upstream or downstream of GNSS-related industries would be suspended.[8] Modern life shows the deep extent to which humanity depends on GNSS and how dangerous it would be if GNSS malfunctioned due to either unintentional or intentional causes.

GNSS was developed for *military* purposes. The downing of Korean Airlines Flight 007 in 1983 made the US decide to open its GNSS to the *civilian* community worldwide.[9] Through several decades of development and with the availability of free

6. NovAtel Inc., *An Introduction to GNSS* (NovAtel Inc., 2010), at 7-9.
7. Scott Madry, *Global Navigation Satellite Systems and Their Applications* (Springer, 2015), at 1.
8. *Ibid.*, at 1-2.
9. *See Statement by Deputy Press Secretary Speakes on the Soviet Attack on a Korean Civilian Airliner*, 16 September 1983, in Ronald Reagan, Public Papers of the Presidents of the United States: Ronald Reagan 1983 (United States Government Printing Office, 1984), at 1294-1295.

open signals and cheap receivers, GNSS has more than demonstrated its value in various rapidly growing fields of application but with the core line of its PNT function.[10] GNSS can primarily be used in all forms of transportation by its positioning and navigation services: space flight, aviation, maritime, rail, road, and mass transit.[11] For instance, GNSS intends to offer seamless satellite navigation services to increase flight safety and efficiency.[12] In addition, GNSS timing service with accuracy to nanoseconds is used in an increasing number of areas, including the synchronisation of cellular networks, electrical power grids, and financial networks. GNSS also plays critical roles in land surveying, law enforcement, emergency response including Search and Rescue (SAR), disaster management, environmental protection, precision agriculture, mining, scientific research, unmanned vehicles systems, etc.

[D] GNSS Performance Parameters

The application scope of GNSS performance parameters should be the system as a whole including satellites in outer space, control facilities on the ground and receivers in users' hands or embedded in transport vehicles, or banking equipment and so on, rather than each component which is supposed to have its own requirements in its own field.

Although existing core GNSSs were not designed to meet civilian aviation performance requirements,[13] since the concept of Performance-Based Navigation (PBN) was introduced, the civilian aviation community has been focusing on the definition and content of performance requirements related to GNSS. GNSS performance parameters in the civil aviation sector now include total Area Navigation (RNAV) system requirements,[14] specific signal-in-space (SIS) requirements,[15] and the requirements at the level of GNSS service (*see* §4.03[E]).[16] All of the requirements are however reflected in the same four aspects as follows:

(i) *Accuracy*, which means the difference or error between actual PNT information and computed PNT information.[17] This parameter requires the PNT information provided by GNSS be sufficiently accurate, with a tolerable error in the process of transmitting from satellites, processing in the control

10. In addition to PNT information or data, certain GNSS like BDS could offer the function of short message communications. *See* CSNO, *supra* note 3.
11. UNOOSA, *supra* note 5.
12. National Coordination Office for Space-Based Positioning, Navigation, and Timing (NOC for Space-Based PNT), *Aviation*, http://www.gps.gov/applications/aviation/, last accessed 3 June 2015.
13. ICAO, *Global Navigation Satellite System (GNSS) Manual*, Second Edition-2013, Doc 9849, AN/457, at 1-2.
14. ICAO, *Performance-Based Navigation (PBN) Manual*, Third Edition-2008, Doc 9613, AN/937, at I-(iii).
15. ICAO, *Aeronautical Telecommunications, Volume I Radionavigations*, Annex 10 to the Convention on International Civil Aviation, Sixth Edition, July 2006, at ATT D-1 to ATT D-7.
16. ICAO, *supra* note 13, at 1-2.
17. *Ibid.*, at 2-1.

segment, correcting by an augmentation system and computing by receivers, so as to transfer true position and right time to intended users.

(ii) *Integrity*, which is a measurement of the veracity of the correctness of PNT information provided by GNSS. This parameter requires GNSS be able to alert users within the allowed time, when GNSS errors reach the designated alert limits and should not be used for the intended operation.[18]

(iii) *Continuity*, which is the capability of GNSS to perform its function without unscheduled interruption during the intended operation.[19] This parameter requires that the loss of GNSS service or signals, whether accidental or unintentional, be less than the designed duration unless the intended users have appropriately been given advance notice.

(iv) *Availability*, which refers to the access to PNT information provided by GNSS in the intended period of time.[20] This parameter requires GNSS service or signals be available for use in the intended duration of operation and, in a broader sense, a certain period must be left to make the alternatives ready before GNSS was turned off to the intended users.

At the core of these performance parameters is the overriding principle of safety,[21] as well as efficiency. Although the four criteria originate from civil aviation, they have already been taken to define the level of safety required for a navigation system in general.[22] Therefore, these criteria are used as standards and tools to check whether PNT information provided by GNSS meets users' needs or providers' guarantees, and to determine whether or not a GNSS provider could be liable or not (*see* §2.04[C]).

[E] Global Nature of GNSS

GNSS is *global* in nature because of its global coverage and worldwide deployment. On the one hand, GNSS provides PNT services or signals to various users, such as ships and aircraft serving international routes with global mobility and international interests as regulated by the Cape Town Convention on International Interests in Mobile Equipment; on the other hand, GNSS ground facilities and augmentation system are globally deployed and constructed and located in different territories, which indicates that various jurisdictions exist in a legal dispute (*see* §2.05).

In addition, the cooperation and joint efforts in terms of interoperability and compatibility between different systems makes GNSS more 'global'. More satellites mean higher accuracy and wider coverage. The interoperability of GNSS multiplies the

18. *Ibid.*
19. *Ibid.*, at 2-2.
20. *Ibid.*
21. ICAO, *Charter on the Rights and Obligations of States Relating to GNSS Services*, Resolution A32-19, the 32nd Session of ICAO Assembly, 1998.
22. OECD, *Space 2030: Tracking Society's Challenges* (OECD publishing, 2005), at 155; *What Is GNSS?*, http://egnos-portal.gsa.europa.eu/discover-egnos/about-egnos/what-gnss, last accessed 13 January 2016.

navigation satellites available for users. At the same time, each GNSS has its own main service area, and the combination of different systems, under the principle of compatibility between combined systems, benefits humanity across the globe with better quality service. For example, GLONASS is developed especially for the use in high latitudes of both the Northern and Southern Hemispheres,[23] but BDS are paying much more attention to middle and lower latitudes.[24] Therefore, China and Russia signed 'China's BeiDou system and Russian GLONASS system Compatibility and Interoperability Cooperation Joint Statement' in 2015,[25] which established a legal basis for the cooperation of the two systems.

The above *technical* global nature leads, in an institutional light, to (i) the international trend of *GNSS governance*, which is why the International Committee on Global Navigation Satellite Systems (ICG) was established under the umbrella of the United Nations (UN) in 2005. The ICG has been contributing itself to promoting the use of GNSS infrastructure on a global basis and to facilitating the exchange of information.[26] In turn, the above technical global nature leads, in a legal light, to (ii) the international characteristics of *GNSS law*, presented by worldwide shareholders, regulatory institutions, certification systems, contractual relationships, and other institutional and legal issues. In particular, the global nature makes GNSS civil liability fairly international but quite complicated, for example, worldwide harmful interference, victims in different jurisdictions, the application of international conventions and bilateral agreements (*see* §2.05).

§1.03 GNSS RISKS AND RISK MANAGEMENT

GNSS presents opportunities for increased safety, security, and efficiency, but all of that must be secured by accuracy, integrity, availability, and continuity requirements without malfunctions, degradations, and harmful interference. Unfortunately, GNSS is now facing technical, financial, institutional, and legal risks which are potential causes of accidents, damage, and detonators of GNSS civil liability (*see* §2.04[C]).

First, GNSS' *technical* vulnerability makes it hard to guarantee the expected service or signals around the clock. While GNSS' inherent error is restricting its accuracy and integrity, the most notable area of concern for users in all sectors, in particular national critical infrastructure,[27] is wilful or inadvertent, lawful or unlawful interference including cyber terrorism, which is now challenging GNSS continuity and availability. In addition, GNSS has its technical limitations. For example, GNSS signals

23. Brian Harvey, *The Rebirth of the Russian Space Program* (Springer, 2007), at 127.
24. *Voice of China: China Is Going to Establish the System of BeiDou Grounded-Based Service, Whose Accuracy Is Better than That of GPS*, in Chinese, http://www.sbsm.gov.cn/article/mtbd/201410/20141000018566.shtml, last accessed 21 June 2015.
25. CSNO, *Sino-Russia Signed a Joint Statement of the Compatibility and Interoperability Cooperation of Beidou and Glonass System*, http://www.beidou.gov.cn/2015/05/11/201505114e742d97c7084dedb4b2024f6c8be64b.html, last accessed 13 January 2016.
26. Someswar, et al. *supra* note 1, at 17.
27. The US Department of Homeland Security (DHS), *Fact Sheet: National Risk Estimate: Risks to U.S. Critical Infrastructure from Global Positioning System Disruptions*, June 2013, https://www.gps.gov/news/2013/06/2013-06-NRE-fact-sheet.pdf, last accessed 13 January 2016.

are unable to penetrate buildings, water, and other obstacles, which makes it rather difficult to construct a seamless satellite navigation environment in a metropolis or in valleys, which would inevitably decrease the level of safety of unmanned vehicles.

Second, most GNSSs are under high *financial* pressure. The development and operation of GNSS need rather substantial investment which is a major entry barrier for most user States.[28] For example, the fully operational phase of Galileo has been postponed repeatedly from 2008 onwards, mainly due to the financial crisis. In addition, short-lived components in particular satellites need to be monitored, maintained, or renewed continuously, otherwise GNSS may become malfunctional and lose its global coverage, which was exactly the real-life scenarios of GLONASS in the 1990s.[29] GNSS providers' current policy is to be free of direct users fees at least for open signals, and this is expected to continue for some time (*see* §5.02[B]), since, to date, no appropriate cost recovery mechanism has been figured out.[30] GNSS investors have little opportunity to get a return on their investment in the near future, let alone potential profit, which led to Galileo's Public-Private Partnership (PPP) approach falling apart in 2006.[31] Although the US government committed to provide GPS Standard Positioning Service (SPS) signals on a continuous worldwide basis and without direct users' fees, this was subject to 'the availability of funds';[32] conversely, this means that, if there are no funds, then there are no promised signals.

Third, GNSSs are facing a series of *institutional* challenges. Except for Galileo, which is dedicated to being the first *civil* GNSS, all other core GNSSs are dual-use but under *military* control. Meanwhile, most user States are relying on extraterritorial GNSS facilities outside their control.[33] Because of the US Selective Availability (SA) policy before 2000 and the modern concept of 'navigation warfare', political or military conflicts may therefore lead GNSS providers to shut down their military and/or civil signals to deny hostile use in the area of military operations.[34] This consideration was the exact motivation for Europe to develop Galileo,[35] and for China to develop BDS.

In addition, GNSS management problems cannot be ignored. Good mechanisms for spectrum allocation are a precondition to ensuring good access to navigation signals. Yet, many conflicts exist here, both internationally and domestically. Two typical examples are the debate between Galileo and BDS on the issue of frequency

28. Ruwantissa Abeyratne, *Space Security Law* (Springer, 2011), at 23.
29. Harvey, *supra* note 23, at 129.
30. Tom Logsdon, *Understanding the Navstar: GPS, GIS, and IVHS* (Springer, 1995), at 178.
31. Glen Gibbons, *European Court of Auditors Lambastes Galileo Satellite Navigation Program*, http://www.insidegnss.com/node/1426, last accessed 26 June 2015.
32. ICAO, *Report on the Establishment of a Legal Framework with Regard to CNS/ATM Systems Including GNSS*, A35-WP/75, LE/5, 28/07/04, at A-17.
33. Caroline Mantl, *Risk Management: The EUROCPNTROL System* (UNIDROIT, 2012), C.D. (91) 6, March 2012, at 4.
34. The US White House, *Fact Sheet: U.S. Space-Based Positioning, Navigation, and Timing Policy*, December 15, 2004, https://www.gps.gov/policy/docs/2004/, last accessed 13 January 2016.
35. Chalmers Johnson, *Nemesis: The Last Days of the American Republic* (Metropolitan Books, 2008), at 235.

overlap,[36] and the spectrum fight between GPS and LightSquared in the US.[37] Concomitantly, incomplete supervision of staff, inadequate training of personnel, unreasonable workflow, and human fault in the operation of software or hardware may cause a GNSS incident or accident.[38]

Fourth, in the case of GNSS failure by any trigger whatsoever, under current *legal* practice and research there exist few guarantees to make innocent victims get fair, prompt and adequate compensation since:

(i) no particular international legislation or related complete proposal for GNSS has been concluded, but meanwhile no clear structure or roadmap to apply either current international conventions or national laws has been created by legal professionals, considering conflicts of different jurisdictions caused by the global nature of GNSS (*see* §1.02[E]).

(ii) the principle of State sovereignty is always a stumbling block leading GNSS civil liability theory into practice (*see* §5.03[B]), as it has been embedded deeply into today's legal system such as Article 1 of the Chicago Convention and Item b) of Article 3 of the Charter on the Rights and Obligations of States Relating to GNSS Services.

(iii) jamming and spoofing devices are strictly prohibited by most national laws but, unfortunately, still widely available on the Internet and easily accessed, which shows the lack of respect for the law concerning harmful interference to GNSS signals.

(iv) the line between the liability for damages caused by GNSS service or signals and that by GNSS value-added or supported service is not always clear for the public, and it is also too hard to be proved and distinguished by the victims without special knowledge, which makes some GNSS providers to undertake the role of popularising law about civil liability issues.[39] The various parties in the GNSS value chain should be distinguished in the context of GNSS civil liability (*see* §2.04[B]).

As the reliance on GNSS continues to grow, the above risks are actual hidden dangers of damage caused by GNSS. Respective measures from technical, financial, institutional, and legal perspectives are proposed as a means of avoiding the approaching disaster.

---

36. Cenan Al-Ekabi (Eds), *Yearbook on Space Policy 2012/2013: Space in a Changing World* (Springer-Verlag Wien, 2015), at 172.
37. NOC for Space-Based PNT, *LightSquared and GPS*, http://www.gps.gov/spectrum/lightsquared/, last accessed 27 July 2015.
38. For example, both GPS and GLONASS have the experiences malfunctioning in 1992 and 2014 respectively. *See* Brandon Ehrhart, *A Technological Dream Turned Legal Nightmare: Potential Liability of the United States under the Federal Tort Claims Act for Operating the Global Positioning System*, 33 Vanderbilt Journal of Transnational Law 2000, at 385; *Glonass Failure Caused by Faulty Software*, http://www.gpsdaily.com/reports/Glonass_Failure_Caused_by_Faulty_Software_999.html, last accessed 17 June 2015.
39. For example, the US has to state that inaccurate map data within GPS devices/apps is not the problem of the government's GPS. *See* NOC for Space-Based PNT, *Help with Address, Route, and Map Problems in GPS Devices and Apps*, http://www.gps.gov/support/user/mapfix/, last accessed 4 October 2015.

For the *technology* community, the central aim is to mitigate GNSS vulnerabilities by anti-risk actions. Technology professionals should devote more attention to the protection of basic signals, considering that no augmentation system could operate without original basic signals. Additionally, to align the remedy with the case above, technical experts should address the following aspects more:

- GNSS monitoring capabilities and appropriate failure indications function, for example, as PBN required in the aviation sector;[40]
- anti-jamming and anti-spoofing technology and detection methods;
- more advanced receivers tracking and catching the strongest signals;
- multi-frequency signal design, and the development of multi-constellation GNSS with more satellites;
- interoperability and compatibility between different GNSSs;
- software protection from cyberattack, and other measures.

For *financial* problems, some fresh vitality should be injected into GNSS investment by the institutional reform on the commercialisation and corporatisation under market mechanisms (*see* §5.03[B][4]). The failure of the PPP of Galileo should not be a stumbling block to searching for new opportunities for cooperation between public funds and commercial investors. Due to the nature of GNSSs' public infrastructures, at the present stage States should keep the public-funds-oriented investment structure. However, the distribution of those public funds must be reasonable among space, ground, and user segments,[41] so as to avoid the 'Buckets effect'.[42] At the same time, the development of applications and the manufacturing of user equipment can be coordinated with the market in the way of a financial support policy without direct investment in industry, like a 'GLONASS Fund' for the GLONASS downstream business chain.[43] Furthermore, a feasible cost recovery mechanism, or pricing policy and revenue model for commercial service should be regulated to maintain the driving force continually for GNSS development. Potential costs of liability should be included in the whole financial budget of GNSS. In addition, taking the successful experience from competitors' cooperation on the International Space Station (ISS) project as a reference, cooperative ventures among GNSS powers will be a potential way to attain cost-savings, although these are not easy to achieve.

From the above scenarios, the author concludes that various technical vulnerabilities, high financial pressures, military-oriented institutional structures, potential management problems, and an insufficient legal system are challenging the expected reliability of GNSS. Although technical development and funding structure reforms

40. ICAO, *supra* note 13, at 2-1.
41. The US Government Accountability Office (GAO), *Global Positioning System: A Comprehensive Assessment of Potential Options and Related Costs Is Needed*, GAO-13-729, at 21.
42. 'Buckets effect' means that the capacity of a bucket depends on the shortest board, which is used by Chinese scholars in social science, and its meaning is similar to 'Liebig's law of the minimum'.
43. *GLONASS Union and VEB Innovations to Promote Navigation Technologies*, http://glonassunion.ru/web/en/pressroom/news/-/asset_publisher/UQppg76eRKAP/content/id/118483, last accessed 18 October 2015.

may mitigate GNSS risks to some extent, those efforts need institutional guarantees and legal safeguards. More importantly, a clear and efficient organisational structure and comprehensive legal system for GNSS could decrease the chance of GNSS malfunctions directly by a more clearly defined supervision framework and increased legal certainty for GNSS stakeholders (*see* below).

§1.04 THE ESTABLISHMENT OF AN INSTITUTIONAL AND LEGAL FRAMEWORK FOR GNSS

[A] The Need for an Integrated Governance of GNSS

Institutional and legal issues of GNSS have been on the agenda since the very beginning of the history of GNSS, and the discussion in this regard reached a climax at the turn of the last century, in particular in the international civil aviation community. Because developments of GLONASS, Galileo and BDS have changed the dynamic into a multinational and multisystem context,[44] and because reliance on GNSS especially in the field of national critical infrastructure continues to grow, it is now the responsibility of States to prepare an integrated governance system consisting of an institutional and a legal framework for the development, operation and exploration of GNSS, with the ambition to mitigate GNSS risks and accidents.

[B] Building the Institutional Structure of GNSS

[1] The Core of Institutional Issues

The core of institutional issues points to the control over availability, continuity, and quality of GNSS service/signals using political influence and/or legal jurisdiction to derive levels-of-safety, liability arrangements, funding and cost recovery mechanisms, management structure, procurement policy, etc.[45] In view of the fact that most States have to rely on the system operated by other States, the aforementioned control should be guaranteed by an institutional framework both at national and international levels. Provider States should ensure that their systems comply with performance parameters required by the users through a systematic internal institutional framework. User States need to increase their indirect control over GNSS outside their territories by a series of international institutional arrangements.

44. Madry, *supra* note 7, at 85.
45. M. J. Asbury, *Some Institutional Factors and Aspects Relating to a Civil Global Navigation Satellite System*, 47(2) Journal of Navigation 1994, at 136.

[2] Internal Institutions

First and foremost, in view of the inherent military interests of most GNSSs (GPS, GLONASS, and BDS), military use of civil GNSS (Galileo),[46] and civilian needs for PNT capability offered by the above systems, the national framework of supervision over GNSS is intended to be composed of both the military and civil sectors, and to give civil interests enough consideration in all decision-making processes. For example, the US President established a joint civil and military body, the US National Executive Committee for Space-Based PNT, for the management of GPS which is operated by the US Air Force.[47] In addition, the framework should be comprehensive enough to include the following or similar institutions:

(i) *Spectrum Management Authority.* As a radio navigation system,[48] the good operation of GNSS is guaranteed by effective spectrum management. This authority needs full competence on domestic assignment of GNSS-related frequencies. In the use of foreign GNSS signals, it may be regulated to require a licence to that end.[49] Moreover, this authority should be granted enough law enforcement capacity so as to achieve institutional control over harmful interference, in particular spoofing and jamming of GNSS signals. This authority should also establish a procedure of cooperation among the regimes of civil liability, State responsibility, and criminal punishment to increase the cost of illegality incurred due to jamming and spoofing.

(ii) *GNSS Operation Entity.* This entity could be either a civil or military authority, while the contractor is supposed to be a company under private law (*see* §5.03[B][4]). This entity is responsible for the deployment, maintenance, and modernisation of satellite constellation and ground facilities. Being against GNSS risks, this entity is advised to back up additional navigation satellites to allow flexibility to launch on demand when the operational need arises in the event that a significant series of failures occur.[50]

(iii) *GNSS Supervisory Authority.* This authority should undertake the role of supervising the operation of GNSS and, at the same time, work as an independent agency to hear claims from GNSS users.

First, this authority should be responsible for managing and monitoring GNSS programme funds, reviewing and approving GNSS development plans,

---

46. The European Parliament voted to agree to the use of Galileo for operations related to the European Security and Defence Policy (ESDP). *See* Karl von Wogau, *On Space and Security*, 2008/2030(INI), Report of 10 June 2008 to European Parliament, A6-0250/2008.
47. NOC for Space-Based PNT, *National Executive Committee for Space-Based Positioning, Navigation, and Timing (PNT)*, http://www.gps.gov/governance/excom/, last accessed 5 November 2015.
48. Durk van Willigen, *Radio Navigation: Perspectives and Challenges*, presented to NAV02 – GNSS Vulnerability, London, 5-7 November 2002.
49. For example, Section 301 of the US Communications Act of 1934.
50. Bernard J. Gruber, et al., *Space Superiority, Down to the Nanosecond: Why the Global Positioning System Remains Essential to Modern Warfare*, 5 Air & Space Power Journal 2013, at 111.

and encouraging research into new technology such as the *GNSS H2020 project*.[51]

Second, this authority should assure the security of both GNSS intangible PNT data and tangible assets[52] through the establishment of a subordinate body like 'Galileo Security Monitoring Centre' under the umbrella of the European GNSS Agency (GSA),[53] which is now working with the 'Security Board for the European GNSS Systems' set up by the European Commission (EC).[54]

Third, this authority should fulfil the duty of information disclosure on *inter alia* health conditions of GNSS satellites and the impact to the users on GNSS performance arisen from approaching space weather.

Last, in light of the ICAO approach and the Single European Sky (SES) Regulation,[55] this authority should establish a subordinate monitoring body, like the 'GNSS monitoring centre for civil aviation of the Russian Federation',[56] to provide a timely warning in case of GNSS malfunctioning or signal failure particularly in the civil aviation sector. This authority should also deal with claims about any damage caused by GNSS and report on GNSS degradations, disruptions, and other anomalies. For example, in the case of GPS, the above role is now taken by the US Coast Guard Navigation Center (NAVCEN) in terms of land-based or maritime PNT signal problems,[57] and by the US Federal Aviation Administration (FAA) for aviation (domestic or international) usage thereof.[58]

(iv) *GNSS Certification Authority*. In the context of GNSS, certification refers to the confirmation by either an internal or an external review body that GNSS complies with the required performance parameters. A certification may not guarantee any feature or functionality of the certified system, signal, service, or product, but it does help mitigate risks in operation and increase the rate that all factors work as intended and planned, especially in safety-critical fields.[59] Also, the dominant power on certification schemes is one of the key

51. GSA, *GNSS H2020 Projects*, http://www.gsa.europa.eu/gnss-h2020-projects, last accessed 24 October 2015.
52. For example, the EU is responsible for all questions relating to the security of EGNOS and Galileo. *See* Article 13 of Regulation (EC) No 683/2008.
53. GSA, *Galileo Security Monitoring Centre*, http://www.gsa.europa.eu/security/gsmc, last accessed 24 October 2015.
54. *See* Commission Decision 2009/334/EC of 20 April 2009 establishing an expert group on the security of the European GNSS systems.
55. EUROCONTROL, *PBN Implementation Issues: Adapting SES Framework to GNSS*, Version 5, 06/09/2013, at 5.
56. Oleg Denisenko, *Proposals on the Development of the International GNSS Monitoring and Assessment System*, presented to Working Group Meeting of ICG, Prague, Czech Republic, 11-13 November 2014.
57. NAVCEN, *GPS Problem Reporting*, http://www.navcen.uscg.gov/?pageName=gpsUserInput, last accessed 24 October 2015.
58. FAA, *GPS Anomaly Reporting Form*, http://www.faa.gov/air_traffic/nas/gps_reports/, last accessed 24 October 2015.
59. Martin Grzebellus, *Is Certification of Galileo a Bureaucratic Overhead?*, 4 Coordinates 2008, at 11-12.

considerations to increase the control over GNSS.[60] The scope of certification should cover the whole system,[61] including signals in space, safety-critical applications and user equipment. Although so far no certification regime exists for the whole system itself, GSA's achievement of 'ISO 9001 Certification' was a great start for the certification of a national or regional GNSS management system.[62] At present, national authorities could focus on the certification of GNSS signals, safety-critical applications and receivers respectively. However, since different domains relying on GNSS are based on different certification requirements for authentication, national authorities have to grant approval following different sets of rules, guidelines, and interests.[63] Therefore, the challenge that national authorities are facing is to figure out respective standards established by various regulatory texts in force; fortunately, however, at least in the civil aviation sector, ICAO Annex 10 Aeronautical Telecommunications and International Telecommunication Union (ITU) recommendations could be taken as references.[64]

(v) *GNSS Sustainable Development Authority*. In order to ensure the continuity of GNSS benefits in the long term, a national governance body needs to consider sustainable development issues appropriately. This authority should be a body of policy and strategy research which works out feasible pathways to drive ongoing GNSS development without break. The scope of research should cover most subjects related to GNSS development but in particular the following aspects:
   - national research project into GNSS risk;[65]
   - exploration of how to promote the commercialisation and corporatisation of GNSS;
   - measures to create a good business environment for the GNSS industry and promote GNSS competitions to increase the quality of service or signals;

60. Johan Lembke, *Competition for Technological Leadership: EU Policy for High Technology* (Edward Elgar Publishing, 2002), at 74.
61. Alessandra A.L. Andrade, *The Global Navigation Satellite System* (Ashgate, 2001), at 120; Recommendation 1 in Attachment 2 to Appendix C. Recommendations of LTEP, *Global Air Navigation Plan*, Third Edition-2007, Doc 9750, AN/963.
62. GSA, *Press Release: GSA Earns ISO 9001 Certification*, GSA/PR/14/09, Prague, 11 December 2014.
63. *International Symposium on Certification of GNSS Systems & Services*, http://www.dgon-cergal.org/index.php?id=23, last accessed 25 October 2015.
64. ICAO, *supra* note 13, at 5-1.
65. For example, *National Risk Estimate: Risks to U.S. Critical Infrastructure from Global Positioning System Disruptions*, launched by the DHS. At the same time, the US GAO conducts certain research into GPS development almost annually, for example, GAO, *GPS: Actions Needed to Address Ground System Development Problems and User Equipment Production Readiness*, Report to Congressional Committees, GAO-15-657, September 2015.

- efforts to reduce excessive reliance on GNSS especially in critical infrastructure sectors, such as the US FAA's Alternative Positioning, Navigation, and Timing research;[66]
- construction backup facilities of GNSS from the EU's experience for the twin control centre of Galileo,[67] and other aspects.

(vi) *International Relationship Office.* After GPS' outstanding start in GNSS decades ago, since the beginning of twenty-first century its 'unipolar moment' has passed with the advent of new global and regional systems. This makes international cooperation among providers increasingly important.[68] At peer-to-peer level, this office should expand international partnerships bilaterally on a wide scale for technology communication, frequency compatibility, system interoperability, installation of facilities in foreign territories,[69] promotion of market access for GNSS technology, devices, signals, and other industries. In addition, GNSS-related standardisation, certification, security, and trade matters are also strongly linked between each other.[70] At a multilateral level, this office, as national representative, would need to forge good connections with GNSS-related international organisations such as ICAO, ITU, and the ICG (*see* §1.04[B][3]).

The above framework would need to be updated with the reform of the GNSS operation structure, such as the commercialisation or corporatisation (*see* §5.03[B][4]). In addition, the above internal institutions are not needed for all States, for instance, the GNSS Operation Entity is only necessary for provider States. Also, certification, operation and other competences could be transferred to the competence of a regional or international organisation (*see* below).

[3] International Organisations

The multi-GNSS era with multi-constellation and multi-frequency is expected to arrive around 2020, which means that international cooperation among *provider States* should be much more addressed than ever. At the same time, the acceptability of an extraterritorial GNSS by *user States* depends on the level of control ensured by *international* legal and institutional arrangements, being aware of the less possibility for provider States to consider the interests of users in essence under their *national*

66. National Research Council, *Global Navigation Satellite Systems: Report of a Joint Workshop of the National Academy of Engineering and the Chinese Academy of Engineering* (Washington, DC: The National Academies Press, 2012), at 127.
67. *Galileo Service Interruption for Ground Segment Upgrade*, http://galileognss.eu/galileo-service-interruption-ground-segment-upgrade/, last accessed 30 November 2015.
68. Madry, *supra* note 7, at 83.
69. *See* Anna Masutti, *Legal Problems Arising from the Installation of the Galileo and EGNOS Ground Stations in Non-EU Countries*, 37 Air & Space Law 2012, at 65.
70. *International GNSS Cooperation*, http://www.gnss.asia/international-gnss-cooperation, last accessed 27 October 2015.

system.[71] Both inter-system cooperation and the level of control required above could be achieved in various stages, including:

- bilateral agreements, such as the 'Agreement on the Promotion, Provision and Use of Galileo and GPS Satellite-Based Navigation Systems and Related Applications';[72]
- multilateral agreements, such as the tripartite agreement between the EC, European Space Agency (ESA), and EUROCONTROL on the European GNSS;[73]
- agreements with an inter-governmental organization (IGO) or by becoming a member of an international organisation such as the ICG, ITU, and International GNSS Service (IGS);
- international regulatory framework including the enforcement of international regulations on GNSS operation or service/signal provision.[74]

Considering a few GNSS providers on the one side, and hundreds of user States on the other side, peer-to-peer relationship between them, as showed by the first stage above, would no longer be suitable. Certain multilateral institutions would be essential for the other three stages, which could be divided into the following three levels:

(i) *Provider level.* The relationship between each GNSS provider should be a win-win by promoting cooperation and decreasing barriers with respect to technical, financial, and commercial aspects, for example, by avoiding trade protectionism and retaliation regarding authorisation for foreign GNSS signals/services.[75] In addition, the expected negotiations and cooperation need an international platform which could gather GNSS providers together to exchange information and, more importantly, promote compatibility and interoperability, with the participation of each 'International Relationship Authority' as proposed above. Fortunately, the establishment of the ICG on December 2005 under the umbrella of the UN filled the gap, which was a key milestone in the history of GNSS. It would be a better choice for GNSS providers to join this international club, in particular its 'Providers' Forum', due to its reputation, prestige and self-esteem with increasing public awareness and industrial actors in the GNSS chain.[76] However, the ICG mainly focuses on technical issues, and was founded on a voluntary basis as an informal body. The above characteristics limit the ICG's competence for

71. Sang Wook Daniel Han, *Global Administrative Law: Global Governance of the Global Positioning System and Galileo*, 14 ILSA Journal of International & Comparative Law 2008, at 591.
72. NOC for Space-Based PNT, *Agreement on the Promotion, Provision and Use of Galileo and GPS Satellite-Based Navigation Systems and Related Applications*, http://www.gps.gov/policy/cooperation/europe/2004/gps-galileo-agreement.pdf, last accessed 29 January 2016.
73. *Agreement Between the European Community, the European Space Agency and the European Organisation for the Safety of Air Navigation on a European Contribution to the Development of a Global Navigation Satellite System (GNSS)*, L194, 10/07/1998, at 16.
74. *Ibid.*
75. *Official: Foreign GNSS Signals Need FCC Authorisation for Use in United States*, http://www.insidegnss.com/node/4334, last accessed 29 October 2015.
76. Han, *supra* note 71, at 581.

making binding decisions, which is why the user community, including States and other organisations such as ICAO and the International Maritime Organization (IMO) showed less interest in it.[77] Still, the ICG would likely be the most suitable body for GNSS cooperation at provider level,[78] with the effort of increasing competence and legal power.

(ii) *User level.* Although users are widely spread throughout the world in every walk of life, for now it is still a seller's market that dominates the provision of GNSS signals. For users, the ideal model is an 'Civil Global Navigation Satellite System',[79] which would be operated by an international legal entity, but which would be supervised by an independent IGO, drawing lessons from the successful commercialisation and corporatisation practice of the International Telecommunications Satellite Organization (Intelsat), the European Telecommunications Satellite Organization (Eutelsat), and others in the telecommunication service market (*see* §5.03[B][4]). This notwithstanding, in the current pattern of the GNSS market there is no indication of owners relinquishing the exclusive ownership of GNSSs.[80] Fortunately, however, the aviation and maritime community, represented by ICAO and IMO respectively, have at least taken certain responses to GNSS to ensure their safety. ICAO initiated negotiation with provider States concerning quality and duration of GNSS in 1991, and two years later ICAO GNSS Panel was established to develop Standards and Recommended Practices (SARPs) in support of aeronautical applications of GNSS (*see* §4.03[E]).[81] The above tasks were achieved by means of the exchange of letters (*see* §4.03[C]), and Volume I-Radio Navigation Aids of ICAO Annex 10 respectively (*see* §4.03[E]). Although IMO acknowledged its inability to fund a worldwide radio navigation system, it accepts current GNSS services through a recognition procedure by the IMO's Maritime Safety Committee (MSC).[82] GPS, GLONASS, and BDS have already been included into IMO's World-Wide Radionavigation System successively.[83] Meanwhile, both ICAO and IMO have been working on a legal framework for GNSS for decades.

However, the scope which ICAO and IMO could influence is limited to aircraft and vessel operators, whereas the needs of the users regarding GNSS in other domains, such as land mobile transport, and financial, electronic

77. *See* member lists and type of the ICG, http://www.unoosa.org/oosa/en/ourwork/icg/members.html, last accessed 29 January 2016.
78. Ingo Baumann, *State of Play in the European Union: Liability for GNSS Signals and Services* (November/December Inside GNSS 2015), at 42.
79. UNOOSA, *Meeting International Responsibilities and Addressing Domestic Needs: Proceedings*, 2006, ST/SPACE/32, Vienna, at 434; IMO, *Maritime Policy for a Future Global Navigation Satellite System (GNSS)*, A20/Res.860, 2 December 1997.
80. Han, *supra* note 71.
81. Jim Nagle, *ICAO Policy on GNSS, GNSS SARPs and Global GNSS Developments*, presented to ICG-2, Bangalore, India, 5-7 September 2007.
82. IMO, *Worldwide Radionavigation System*, Resolution A.1046(27), A 27/Res.1046, 20 December 2011.
83. *See* IMO, SN/Circ.182, SN/Circ.187 and SN.1/Circ.329 respectively.

and communication systems, should be identified and protected within a wider international supervision framework for a long-term solution, as discussed below.

(iii) *Supervision level.* Complex technology coupled with a market of oligopoly results in an unequal power relationship between legal equals - GNSS providers and users. It is therefore necessary for international governance bodies to intervene, for the user, and hold the provider accountable.[84] *Supervision*, rather than *cooperation*, is the most powerful tool to reach the said purpose, in addition to the control of GNSS *ownership*, even though, with regard to the latter, there is no feasible way for it to change in the near future. Meanwhile, both GNSS providers and users need to find protection together against internal and external signal interference at a global coverage level. Therefore, a global administrative institution with a certain competence over both providers and users is needed. Considering conflicts of interest, in practice, of GNSS competitors, this institution could not be designed and achieved easily just by transferring and combining a governance structure available in each domestic setting to the global setting. The institution should have fiscal and supervisory accountability to its Member States, and the top management should be subject to electoral accountability to such States.[85] This institution could be established as a new international body, but it could also be achieved by expanding the competence of the ICG from an informal body to a regular body or subordinate part of a future specialised agency governing space issues. For example, if ICAO were developed for space governance in the future, the ICG could also consider being one of its departments for matters of GNSS.[86] But both of the above paths have quite a long way to go before becoming a reality. Before that time, the GNSS community could continue with the *status quo*, i.e., having separate governance with ICAO for GNSS affairs in the civil aviation sector, and IMO for maritime navigation. Regardless of the path that is selected, the following aspects must be however considered carefully in the whole supervision framework:

(i) *Radio frequency.* Under the current global information and communication technology (ICT) governance framework, ITU, in particular its *Radiocommunication Sector* (ITU-R), is, *inter alia*, responsible for frequency allocation and allotment, and elimination of harmful interference globally. Although strictly speaking the relationship between 'Radiocommunication' and 'Radionavigation' is more akin to that of a 'brotherhood' than that of a 'father and son', the frequency used by GNSS has to abide by ITU's management and regulations, which is already an established

84. Han, *supra* note 71.
85. *Ibid.*
86. *See* Ram S. Jakhu, et al. (Eds), *The Need for an Integrated Regulatory Regime for Aviation and Space: ICAO for Space?* (Springer-Verlag/Wien, 2011), at 131.

fact. Nevertheless, ITU should play a stronger role in settling disputes about GNSS-related frequency, as discussed in (d), below.

(ii) *International certification and standards*. Except for the above-mentioned acceptance and recognition procedure by IMO, which showed the prototype of certification, at present no international certification system is available for the GNSS system itself, signals in space, safety-critical applications, and user equipment. Already in the year 1998, ICAO formulated detailed recommendations on GNSS certification.[87] Since then, however, no further progress has been achieved. Therefore, the efforts should be continued by the GNSS community to establish or find a proper institution for certification issues at an international level. The proposed institution would need to be independent and professional, since it would deal with highly technical matters concerning safety of life (SoL).[88] For the purpose of authentication, the proposed institution should also be entrusted by with the right to obligate GNSS providers to submit their performance report as requested. Consequently, there is a need to formulate a series of globally recognised performance benchmarks in consultation with other international organisations, such as those dealing with air and maritime navigation, and relevant national authorities.[89] Before the above is achieved, there could be proposed interagency cooperation among IGOs, including ICAO, IMO, ITU, as well as non-governmental organisations (NGOs) such as the International Organization for Standardization (ISO) on GNSS certification and international standards.[90] Meanwhile, since no conflicts are perceptibly anticipated between international certification and domestic certification, the efforts on certification by national GNSS certification agencies should always be continued.

(iii) *Monitoring and transparency*. Although each provider State is recommended to establish its own internal monitoring framework (*see* §1.04[B][2]), its credibility would never match that offered by an international independent body because of the internal political situation. In addition, public transparency, or at least interagency transparency provides monitoring information to the public, which is necessary for evaluating performance, taking remedial actions, and promoting peer, reputational, and market pressure,[91] and this could help user States increase the sense of control over GNSS to some extent. Actually, the job of both GNSS monitoring and transparency could be performed by

87. ICAO, *Global Air Navigation Plan*, Third Edition-2007, Doc 9750, AN/963, at App C-8 to App C-9.
88. Han, *supra* note 71.
89. Andrew Wilson (Eds), *Galileo: The European Programme for Global Navigation Services* (ESA Publication Division, 2005), at 21.
90. ISO is also working on the standards in GNSS field, such as the document of 'ISO 17123-8:2015'. For more information, please visit http://www.iso.org/iso/home/store/catalogue_ics/catalogue_detail_ics.htm?csnumber=62961, last accessed 05 February 2016.
91. Han, *supra* note 71.

potential international certification organisations. Alternatively, the ICG, which established a subgroup on 'International GNSS Monitoring and Assessment (iGMAS)', could link the existing GNSS monitoring centres to a new ICG portal with unified parameters, which would allow GNSS users worldwide to easily access GNSS monitoring information and products by simply checking the ICG webpage.[92] Furthermore, in the process of prompting an international GNSS monitoring programme, the role of the IGS, a voluntary federation of over 200 self-funding GNSS-related bodies, should be considered seriously.[93]

(iv) *Dispute settlement/sanctions*. International disputes are inevitable in global expansion, competition, and cooperation of GNSS, such as: frequency overlapping between BDS and Galileo;[94] cyber war on jamming GPS signals against South Korea from North Korea;[95] tensions between the US and the Russian Federation on the construction of ground stations;[96] GNSS patent disputes between the UK and the US;[97] potential navigation satellites crashing in space; and signal or service trade barriers. Therefore, an international co-ordinating framework, or more powerful sanctions are essential for the purpose of GNSS dispute settlement. For the States' conflicts on GNSS frequency, including harmful interference, ITU plays an important role in coordination,[98] even though its procedure is limited by its soft power on enforcement and binding effect.[99] In addition, other types of disputes could be resolved under each specific framework, such as the World Trade Organization (WTO) for GNSS signal/service trade barriers. Nevertheless, a coordination procedure and serious sanctions arrangements should be on the agenda of any organisation whatsoever so as to avoid a navigation war, i.e., a military attack on enemy dual/civil-use GNSS, in particular navigation satellites, which would be a disaster for civilian users.

Despite the internal or international arrangement of institutional control over GNSS above, its purpose is always to achieve the balance of interests between providers and users, and more importantly, to ensure public security and safety.

92. Denisenko, *supra* note 56.
93. Urs Hugentobler, et al., *IGS Activities on GLONASS*, presented to GGOS-RAS Meeting, Vienna, 9 April 2013.
94. Al-Ekabi, *supra* note 36.
95. Han, *supra* note 71.
96. *U.S. Still Not Allowing GLONASS Stations*, http://gpsworld.com/us-still-not-allowing-glonass-stations/, last accessed 31 October 2015.
97. Dee Ann Divis, *UK Revokes Key GNSS Patent That Sparked Dispute over Cooperation, Interoperability*, http://www.insidegnss.com/node/3253, last accessed 31 October 2015.
98. ITU, *Coordination Procedures*, http://www.itu.int/en/ITU-R/terrestrial/fmd/Pages/coordination.aspx, last accessed 31 October 2015.
99. Chunli Xia, *Study of the ITU Spatial Orbit-Spectrum Resources Allocation and Coordination Rules*, 6 Journal of Beijing Institute of Technology (Social Sciences Edition) 2011, in Chinese, at 92.

[C] Shaping a Legal System of GNSS

The institutional issues above focused on the structure of the governance of GNSS to determine what institutions should be established or involved, and what their respective responsibilities are at both an international and a national level. However, GNSS legal issues are more related to the code of conduct, as well as rights and obligations, of GNSS stakeholders, which include:

- GNSS regulators: the supervisory authorities in both provider States and user States;
- actors from the supply side: owners, operators, and direct providers regardless of their military or civil nature;
- actors from the demand side: direct users and value-added service providers;
- insurers;
- third parties: victims of GNSS damage, perpetrators of harmful interference, etc.

The respective legal relationship among the parties above are quite complicated because they have to be reviewed from three different perspectives: administrative law, criminal law, and civil law (*see* Figure 1.3).[100]

100. Instruction:

A *Criminal Law Relationships*
A1: criminal offence of supervision negligence over GNSS activities with serious loss of life and property.
A2, A3, A4, A5: intentional or unintentional offences because of GNSS-related activities.

B *Administrative Law Relationships*
B0: global administrative governance relationship
B2, B3, B4, B7: State supervision over GNSS-related activities.
B1, B5, B6: State compensation liability for supervision negligence.

C *Civil Law Relationships*
C0: insurance contractual relationship with correspond insured parties.
C2: competition disputes, patent disputes, etc.
C1-C16: civil liability compensation due to GNSS malfunction, harmful interference, and privacy leak, liability distribution, recourse and so on, which will be discussed in following chapters in detail.

Figure 1.3 GNSS Legal Relationship

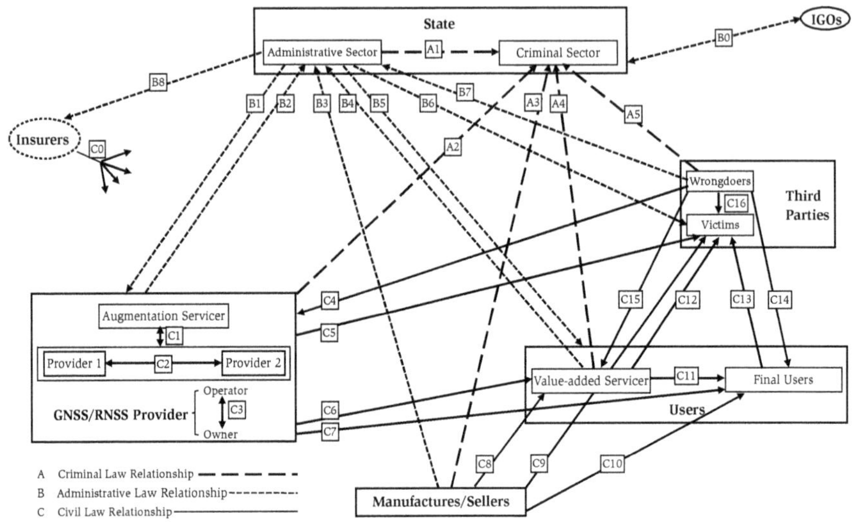

Considering that the issues such as supervision, certification and settlement of disputes have been more or less discussed as institutional problems (*see* §1.04[B]), the following part will only discuss *purely* legal aspects of GNSS in general, whereas some of the legal parameters will be analysed in greater detail in the following chapters:

(i) *The adequacy of the GNSS legal framework.* Could the *status quo* of a framework resolve GNSS' legal problems? In view of this question, the civil aviation community carried out significant work to determine, for instance, whether the 'Statement of ICAO Policy on CNS/ATM Systems Implementation and Operation',[101] the 'Charter on the Rights and Obligation of States relating to GNSS Services', and the 'Development and Elaboration of an Appropriate Long-Term Legal Framework to Govern the Implementation of GNSS' could serve as parts of a legal framework for GNSS (*see* §4.03).[102] In addition, ICAO exchanged letters with both the US and Russia on GNSS service provision (*see* §4.03[C]).[103] Based on the above work, the main GNSS providers, in particular the US, and most user States, in particular developing countries, engaged in a heated discussion: the former insisted that the existing legal framework including the above documents and the system of the Convention on International Civil Aviation (Chicago Convention) was

101. ICAO, *Statement of ICAO Policy on CNS/ATM Systems Implementation and Operation*, ICAO Doc. WP/3-2, LC/29, 28 March 1994.
102. ICAO, *supra* note 87, at App C-6 and App C-13.
103. ICAO, *supra* note 32, at A-17 and A-23.

sufficient,[104] whereas the latter argued that a new long-term legal instrument regarding GNSS services, in the form of an international convention or a contractual framework containing enforceable provisions, is needed.[105] Regrettably, to date no consensus for change has been reached. Yet, considering that the seller's market is being challenged with more GNSSs and RNSSs being developed, it is reasonable to expect the above argument by user States to be restarted with a view to becoming ready for the post-2020 GNSS service. Therefore, Chapter 3 and Chapter 4 will check whether there are enough solutions to the disputes arising from GNSS damages in current international air and space law.

(ii) *Ownership of GNSS*. Generally speaking, ownership is a right to immediate exclusive control of property within a limit-line set by the rule of law.[106] In other words, the notion of ownership is inherently composed of both rights, and obligations or responsibilities at the same time.[107] While on the one hand GNSS owners have the right to make their own decisions on system operation, disposal of property, conclusion of contracts, taking measures to recovery costs, and so on; on the other, GNSS owners are also obligated to implement activities according to law, in particular to ensure the level of safety required by performance parameters. This notwithstanding, the owner, operator, law-executor, and legislators of current GNSS are actually combined in one single party: for GPS, GLONASS, and BDS, it is the provider States represented by their governments; for Galileo, it is the EU represented by the EC. This combination of the role of regulatory body and regulated body means that a provider State's control over its GNSS is too strong for it to be trusted with its own oversight if no external oversight agency exists, and this, after all, is the main concern of user States. Due to a lack of an international balance of a legal framework, the rights and responsibilities of the parties are to be determined, to a large extent, according to the bargaining strengths of those involved,[108] consequently leading to an unequal contest between equal sovereign powers: provider States versus user States. Therefore, a powerful international instrument for GNSS is needed. The commercialisation and corporatisation of GNSS, which could separate the role of the provider from the regulator, has to be addressed (*see* §5.03[B][4]).

(iii) *State responsibility and sovereign rights*. State sovereignty is composed of *internal* sovereignty and *external* sovereignty: the former means that a State has full powers and also responsibility for maintaining legal orders inside its

104. ICAO, *Assembly Resolution on a Practical Way Forward on Legal and Institutional Aspects of CNS/ATM*, presented by the United States of America to the ICAO 35th Session Assembly, ICAO Doc. A35-WP/216, LE/18, 28/9/04.
105. *Legal Aspects of GNSS*, AN-Conf/11-WP/143, 18/9/13, presented by African States to Eleventh Air Navigation Conference, Montreal, 22 September to 3 October 2003.
106. Ben McFarlane, *The Structure of Property Law* (Bloomsbury Publishing, 2008), at 140.
107. Gregory S. Alexander, *Ownership and Obligations: The Human Flourishing Theory of Property*, 43 Hong Kong Law Journal 2013, at 451.
108. *Supra* note 105.

territory; the latter refers to a State being independent of others and defending any meddling in its internal affairs from the outside world.[109] In the context of GNSS, *internal* sovereignty is discussed as the regulation of GNSS matters inside a State's borders; whereas *external* sovereignty, which is related to mutual relationships between GNSS States under international law, is addressed by user States and provider States from the following two different perspectives:

(i) user States are reluctant to let the implementation of out-of-their-control GNSS reduce any degree of statehood on complete dominant power over their critical PNT infrastructure, for example, legislative powers over their airspace confirmed by Article 1 of the Chicago Convention in the provision of Air Navigation Services. In addition, user States expect to accept GNSS services under the non-discrimination principle laid down by Article 2 of the 'Charter on the Rights and Obligations of States relating to GNSS Services';

(ii) provider States are more concerned with immunity from jurisdiction or execution in GNSS-related disputes. One goal of provider States or the EU is to ensure privileges and immunities for their ground stations and related technical and administrative delegation in other States' territories,[110] and another goal, based on *par in parem non habet imperium*, is trying to exclude themselves from all legal proceedings initiated by other States, including claims for compensation for damages caused by GNSS (*see* §5.03[B][2]). User States are fighting for an equal legal arrangement on protection for their sovereignty rights and against provider State immunity in the provision of GNSS services/signals (*see* §5.03[B][3] and §5.03[B][4]).

(iv) *GNSS Liability*. The aforementioned GNSS risks indicate the possibility of GNSS malfunctions regardless of natural or human intentional or unintentional reasons, which may lead to the breakdown of national critical infrastructure relying on GNSS. When any of the above damages occurs, a series of liability issues may emerge according to the degree of participation of *public powers* at three different levels as follows:

(i) criminal liability, triggered by the heavy loss brought about by GNSS-related accidents (*see* §2.03[B]);

(ii) administrative liability, arising from misadministration or regulatory failures over GNSS activities (*see* §2.03[B]);

(iii) civil liability, caused by the failure of GNSS signal/service (*see* below).

In the case of damage caused by GNSS, the civil liability regime, including the insurance system, is intended to ensure that those innocent victims get fair, prompt, and adequate compensation, which is the *heart* of the GNSS legal framework. Similar to the 'The adequacy of GNSS legal framework'

---

109. Jaap Hage, et al. (Eds), *Introduction to Law* (Springer International Publishing Switzerland, 2014), at 159-162.
110. Masutti, *supra* note 69.

discussed above, the civil aviation community on this issue was divided into two parties, which insist on:

(i) continuing the *status quo* by current international legal documents and national legislation;

(ii) new liability arrangement by means of an international convention or a contractual framework.

As a result of unequal political powers, in ICAO, between provider States, namely the US, Russia, China and European States, and user States, most of which are developing States in Africa and Asia, no binding resolutions regarding GNSS civil liability have been concluded. Thus, GNSS providers may freely ignore any calls for a guarantee of liability. This should not however be the final point of the research and practice of GNSS civil liability, since the legal community should be prepared in advance for future massive applications of GNSS, in particular in the safety-of-life domain, which will increase the rate of damages caused by GNSS. Theoretically, the GNSS civil liability regime should cover the following elements, which are also the main contents of the following chapters:

- definition of GNSS civil liability including liable parties and the term 'damage' (*see* Chapter 2);
- current legal framework particularly at an international level (*see* Chapters 3 and 4);
- the fairness of GNSS civil liability (*see* §5.02);
- feasible road path for an acceptable international solution on GNSS civil liability, including the way against the doctrine of sovereign immunity and the feasible choice of legal basis of GNSS civil liability (*see* §5.03).

(v) *Privacy protection*. GNSS can be used to track individuals, including children and criminal offenders, and questions regarding human rights and privacy can arise at the same time. Usually, there are two ways in which the tracker may use to get the location information of a target:

(i) the tracker directly attaches a GNSS unit to the property of the target, such as bags and vehicles; or

(ii) the tracker obtains GNSS location information from a third-party service provider, including GNSS service/signal provider or operator.

From a civil law perspective, the GNSS civil liability issue above-mentioned could be discussed, based on fault liability, as follows:[111] in the former case, the GNSS provider is not liable unless it knows or should know its signals are used to infringe the target's privacy; in the latter case, the GNSS provider is liable if the location information is provided by it illegally. In addition, some administrative and criminal issues could also arise because the trackers could be criminals, even though they could also be law enforcement officials. It is clear that criminals should be sent into prison for

111. The invasion of privacy usually applies general tort liability regime based on fault, such as Article 823 of the German Civil Code; Article 9 of the French Civil Code; Article 2 of the Tort Law of the People's Republic of China.

obtaining a third party's location information illegally. However, one question has not been answered clearly to date, that is whether it complies with the law, in particular rules of evidence, if law enforcement officials get a suspect's location information for investigation purposes but without a warrant. Although this book is not going to touch upon this said issue of warrant further, there are already many cases which can be reviewed in some national laws.[112]

(vi) *Other elements*. In addition to the issues above, many other legal aspects related to GNSS also *need* to be further investigated. Although these aspects fall outside the scope of this book, the following is a short list for future study:
- trade law framework on export and import control of GNSS products, including signals/services themselves and equipment, with reference to the 'WTO Telecom Agreement' of the late 1990s;[113]
- rule on market access and competition order of GNSS providers, operators, and related enterprises;
- intellectual property regulations regarding GNSS technology;
- PNT data business rule, in particular cross-border flow in the ear of Big Data;[114]
- legal arrangement to promote the uniformity of GNSS product standards, system compatibility, and interoperability;
- regulations of issuing a licence for GNSS operation referring to the requirement of remote sensing system operation;[115]
- safety and security of GNSS in a space/navigation war. The operation of GNSS on civil and military functions should be separated, yet the 'law of war' is meant to ban military attacks on civil GNSS facilities and military interference to civil GNSS signals regardless of the military nature of GNSS.[116]

---

112. For example, *United States v. Garcia* (2007) 434 F.3d 994, cert. denied, U.S. Suat Ct. 1 October 2007; *Garcia, Kyllo v. United States* (2001) 533 U.S. 27; 121 S. Ct. 2038; 150 L. Ed. 2d 94, *etc. See* Francis Lyall & Paul B. Larsen, *Space Law. A Treatise* (Ashgate, 2009), at 390.
113. Dee Ann Divis, *Officials Delay First GNSS Authorisation Request; Light-Squared Tries to Leverage Issue*, http://www.insidegnss.com/node/4585, last accessed 9 February 2016.
114. For example, the Directive 95/46/EC of the European Parliament and of the Council of 24 October 1995 on the protection of individuals with regard to the processing of personal data and on the free movement of such data.
115. Article 5 of the Remote Sensing Space Systems Act of Canada reads as follows:

> No person shall operate a remote sensing space system in any manner, directly or indirectly, except under the authority of a licence.

116. The term 'law of war', used in this sentence, is also known as the international humanitarian law or the law of armed conflict, and "it is a set of rules which seek, for humanitarian reasons, to limit the effects of armed conflict." It protects civilians and civil systems, and it also restricts the means and methods of warfare. For example, Articles 48 and 52, the Protocol additional to the Geneva Conventions of 12 August 1949, and relating to the protection of victims of international armed conflicts (Protocol I)', adopted at Geneva on 8 June 1977. *See* International

[D] Concluding Remarks

PNT information provided by GNSS has penetrated into every corner of modern society. Recognising the national security and economic interests brought about by GNSS, the US, Russia, China, and the EU are in various stages of deploying their own GNSSs. GNSS is a generic term of navigation satellite systems, as well as augmented systems thereof, with a global coverage. Being composed of space satellites, ground facilities and user receivers, GNSS is so complicated from a technical point of view that certain performance parameters must be followed in its development and operation, i.e., accuracy, integrity, continuity, and availability.

Nevertheless, GNSSs are now facing technical, financial, institutional, and legal risks at the same time, which could potentially cause accidents and damages to safety-of-life sectors. Innovative solutions to secure the safety of GNSS should be developed to align the remedy with the situation from the above four perspectives respectively. More advanced technology which can transmit signals with better quality and against harmful interference will soon arrive with the endeavours of scientists and engineers. An acceptable cost recovery mechanism of GNSS is expected to be created to decrease financial pressures on GNSS providers. In addition, a domestic management system and an international institutional control over the matters of spectrum coordinates, certification systems, malfunction monitoring, and disputes settlement are essential to balance the interests between user States and provider States in the area of multi-GNSS but with a certain degree of monopoly.

All the solutions proposed above need to be assigned legal certainty, but whether or not the current legal framework is adequate for the legal problems of GNSS still needs to be further researched and developed. GNSS malfunctions and the invasion of privacy through GNSS may trigger liability under civil law as well as administrative law and criminal law. The inherent technical global nature and State-ownership plus State-operation of GNSS however increase the difficulty of guaranteeing the victims' fair, prompt, and adequate compensation for damages caused by GNSS. To focus on the issue of GNSS civil liability, the first things that we need to determine are *who* are the liable ones, *why*, and *how*. Therefore, the next chapter will clearly define the term 'Civil Liability in the Context of GNSS (GNSS civil liability)'.

Committee of the Red Cross (ICRC), *What Is International Humanitarian Law?*, https://www.icrc.org/eng/assets/files/other/what_is_ihl.pdf, last accessed 9 February 2016.

CHAPTER 2

Defining Civil Liability in the Context of GNSS*

§2.01 INTRODUCTION

Concepts are the base and tools for furthering understanding.[1] In this book, the two most important key words are *GNSS* and *civil liability*. Since the technical concept and general background of *GNSS* were explained in the previous chapter, it is reasonable to expect that this chapter carefully defines, from a more legal perspective, the term *civil liability* in the context of GNSS. Civil liability is a term used in numerous national legislations and international conventions, even though there remains to be seen the question as to whether the combination of *GNSS* and *civil liability* differs to the extent of creating a new paradigm compared with the civil liability in the context of traditional navigation error, the carriage of dangerous goods, and environmental damage.[2]

Therefore, the principal objective of this chapter is to define the term *GNSS civil liability* by clearly outlining what kind of GNSS could be qualified as the origin of GNSS damage (*see* §2.02) and, after the general introduction of *liability* in the context of GNSS (*see* §2.03), how to establish *civil liability* at issue (*see* §2.04). In addition, considering the technical global nature of GNSS as mentioned above, this chapter continues to discuss how that factor is transferred to GNSS civil liability in a legal sense (*see* §2.05). The chapter then closes with a few concluding remarks (*see* §2.06).

* Part of this chapter was published as: Dejian Kong, *Civil Liability for Damages Caused by Global Navigation Satellite System: A Conceptual Analysis*, XLI Annals of Air and Space Law 2016, at 313-335.

1. William Michael Reisman, *Soft Law and Law Jobs*, 2(1) Journal of International Dispute Settlement 2011, at 25.
2. *See* Michael Milde, *Air Navigation and Safety: Institutional and Legal Problems of the Global Navigation Satellite System*, IV Temas de aviacion comercial y derecho aeronautico y espacial 2000, at 134.

§2.02 SCOPE OF GNSS UNDER THE TERM GNSS CIVIL LIABILITY

Although the terms GNSS, RNSS, and augmentation system are closely interconnected (*see* Figure 1.1), how well can they be included in the concept of GNSS civil liability? Furthermore, since many extended or value-added PNT services based on GNSS, ranging from Google Maps to GPS tracking services, are playing an increasing role in modern life, can we resolve civil liability disputes arising from those extended services under the framework of GNSS civil liability?

Technically speaking, only a satellite navigation system with global rather than regional coverage can be defined as GNSS.[3] RNSS is technically independent of GNSS with its standalone system constitution and with the transmission of its own signals. However, legally speaking, RNSS and GNSS have no essential difference in terms of civil liability issues. In fact, the major difference between GNSS and RNSS is the coverage: the former is global and the latter is regional. However, even for RNSS, its scope of service usually covers more than one State, which is also a form of *international.* This means that RNSS legal issues are involved in different jurisdictions as well, and this is the fundamental feature of GNSS civil liability (*see* §2.05). Therefore, in this book, the term GNSS civil liability usually includes liability for damage caused by RNSS; the term GNSS civil liability however excludes national navigation satellite system (NNSS) as no *global* or *international* factor exists in NNSS, and it is usually only related to domestic legal systems and has nothing to do with international law.

Similarly, GNSS can provide PNT data alone without any augmentation system. From the perspective of providers, the role of an augmentation system is to aid GNSS in improving its accuracy, integrity, and availability and it is not inherently part of GNSS itself.[4] An augmentation system is only a *processor* or *corrector* of basic signal from GNSS, rather than a *producer* of PNT data, therefore from the standpoint of users such as European Organisation for the Safety of Air Navigation (EUROCONTROL), GNSSs are divided into core constellation systems, including GPS, GLONASS, Galileo, and BDS, and augmentation systems, including Aircraft-Based Augmentation System (ABAS), SBAS, and GBAS.[5] ICAO also confirms this in its definition of GNSS and introduction of GNSS elements.[6] In addition, the augmentation system EGNOS was developed as a precursor and integral part of Galileo,[7] both of which are now part of the European GNSS programmes managed by the EU;[8] the Russian GLONASS SDCM

3. Gelli Manoj Someswar, et al., *Global Navigation Satellite Systems and Their Applications*, 17(1) International Journal of Software and Web Sciences 2013, at 18.
4. NCO for Space-Based PNT, *Augmentation Systems*, http://www.gps.gov/systems/augmentations/, last accessed 19 May 2015.
5. EUROCONTROL, *GNSS*, https://www.eurocontrol.int/articles/gnss, last accessed 19 May 2015.
6. Annex 10 to the Convention on International Civil Aviation, *Aeronautical Telecommunications, Volume I Radionavigations*, Sixth Edition, July 2006, at 3-58 to 3-59; ICAO, *Global Navigation Satellite System (GNSS) Manual*, Second Edition-2013, Doc 9849, AN/457, at 1-1.
7. *See* Communication from the Commission to the European Parliament and the Council: State of Progress of the Galileo Programme, COM/2002/0518 final, 2002/C 248/02, 15 October 2002.
8. GSA, *About EGNOS*, http://egnos-portal.gsa.europa.eu/discover-egnos/about-egnos, last accessed 31 May 2015.

(System for Differential Corrections and Monitoring) is being developed as a component of GLONASS.[9] Therefore, an augmentation system is one of the basic parts of GNSS, and it hence falls under the concept of GNSS civil liability.

However, the liability of a value-added service does not usually belong to the concept of GNSS civil liability. The term 'value-added service' refers to an extended downstream service or application based on GNSS signals in a manner intended to provide additional utility or benefit to the user. This term however excludes the quality improvement, in particular the performance and features of a GNSS signal itself, which is actually done through the augmentation of GNSS.[10] Nowadays, besides the Air Navigation Service (ANS) based on GNSS, in particular the Communications, Navigation and Surveillance/Air Traffic Management (CNS/ATM) being promoted in civil aviation, a typical example of a GNSS value-added service is the navigation service provided by Google Maps or its competitors. Unfortunately, news about what are called 'navigation problems' has been reported frequently,[11] even though most of those problems were caused by map and route planning errors. What GNSS transmits to the receivers is only the signals that contain PNT information, rather than map data, therefore the liable parties in those cases are *map providers*, rather than GNSS providers. If the damage in the navigation were caused by the loss, degradation, and defect of GNSS signals, in principle it would be the GNSS providers which should be held liable thereto. The same applies to other extended services or applications based on GNSS.[12]

Although according to the constitution of GNSS, user equipment is technically speaking an essential segment of GNSS (*see* §1.02[B]), liability arising from the malfunction of user equipment is legally and generally speaking not GNSS civil liability. First, as an end product the receiver receives GNSS signals which then computes its location on its own. If the PNT data offered by GNSS complies with the requirements but the receiver, rather than GNSS provider, computes the data wrongly, the defective receiver or software therein may apply to general rules of tort law or product liability in particular. Second, the space segment and the control segment are controlled and operated by GNSS providers, but the production and sales process of user equipment are conducted by other legal entities; in other words, GNSS providers are not the manufacturer, owner, and controller of GNSS user equipment. This has

9. Sergey Karutin, *System for Differential Correction and Monitoring Updated*, Proceedings of the twenty-fourth International Technical Meeting of the Satellite Division of the Institute of Navigation (ION GNSS 2011), Portland, OR, September 2011, at 1562; Cyrille Boulanger, et al., *Receiver Inter System Bias Impact on SBAS Dual Constellation Positioning and Integrity*, Proceedings of the 26th International Technical Meeting of The Satellite Division of the Institute of Navigation (ION GNSS+ 2013), Nashville, TN, September 2013, at 854.
10. *See* Article 2(q) of the Agreement on the Promotion, Provision and Use of Galileo and GPS Satellite-Based Navigation Systems and Related Applications.
11. *See* Robert Wabash, *9 Car Accidents Caused by Google Maps & GPS*, http://www.ranker.com/list/9-car-accidents-caused-by-google-maps-and-gps/robert-wabash, last accessed 12 January 2016.
12. For a list of value-added service, please visit http://www.gsa.europa.eu/value-added-services-providers, last accessed 12 January 2016.

been confirmed by Article 2.2 of Regulation (EU) No 1285/2013, which reads as follows:

> The system established under the Galileo programme shall be a civil system under civil control and an autonomous global navigation satellite system (GNSS) infrastructure consisting of a constellation of satellites and a global network of ground stations.

At least for the European GNSS programme, the user segment is not addressed directly by the EU; it should be the job of private innovators. Therefore, it is unfair to make GNSS providers responsible for those situations that are completely out of their control. If, however, the receivers are malfunctioning fully or partly because of wrong technical information released by such documents as the 'Interface Control Document' provided by GNSS providers,[13] then the issuing body should be liable. Moreover, if the malfunctioning user equipment has been certified or qualified by a GNSS provider or operator, the latter has the responsibility to guarantee the quality of those devices and undertake relevant civil compensation for damages arising from the malfunctioning equipment.

In conclusion, GNSS civil liability includes liability trigged by core GNSSs, augmentation systems and regional systems, but excludes the one trigged by GNSS value-added services and the malfunction of user equipment. Regardless of the terminal body which caused an accident, as long as the accident is fully or partly resulting from the provision of a *GNSS signal* itself, the concept of GNSS civil liability should and would be applicable.

§2.03 CONCEPT OF LIABILITY IN THE CONTEXT OF GNSS

[A] Responsibility Versus Liability

Responsibility and *liability* are twin legal concepts in general with dual significance. Although there commonly exists an explanation of the difference and similarity between responsibility and liability,[14] and although the two terms have the same single translation in many authentic treaty languages, for example 'responsabilité' in French, 'responsabilidad' in Spanish and '责 任' in Chinese,[15] a clear distinction needs to be made in the legal sense.[16] Responsibility refers to the legal obligation to fulfil a legal duty imposed by law to take care of something, prior to the situation of liability; liability is the negative effect if one did not fulfil his or her responsibility appropriately.

13. For example, CSNO, *BeiDou Navigation Satellite System Signal in Space Interface Control Document Open Service Signal (Version 2.0)*, December 2013.
14. *See*, e.g., Bryan A. Garner (Eds), *Black's Law Dictionary* (WEST, 2009), at 1427.
15. Stephan Hobe, Bernhard Schmidt-Tedd & Kai-Uwe Schrogl (Eds), *Cologne Commentary on Space Law: Volume I* (Carl Heymanns Verlag, 2009), at 104.
16. Liability is a legal term, but responsibility is viewed as a norm in general whether moral, legal, religious, political or any other. Bin Cheng, *Article VI of the 1967 Space Treaty Revisited: 'International Responsibility', 'National Activities', and the 'the Appropriate State'*, 26 Journal of Space Law 1998, at 9.

The distinction between responsibility and liability may be understood much clearer as follows:

> The English legal term "responsibility" means the accountability for a primary obligation of conduct. It is the legal situation which precedes liability, whereas "liability" is attached to a person for the injury resulting from the non-fulfilment of that primary obligation for which he was responsible. He will be subjected to a civil or criminal sanction.[17]

In international law, the demarcation between responsibility and liability is however not always as clear as above, in particular in relation to the regime of obligation in the respect of the injurious consequences arising out of acts not prohibited by international law.[18] The draft articles on the 'Responsibility of States for Internationally Wrongful Acts', approved by the International Law Commission (ILC), defines the 'responsibility of a State for its internationally wrongful acts' as "every internationally wrongful act of a State entails the international responsibility of that State."[19] The legal consequence of an internationally wrongful act includes 'reparation, cessation and non-repetition as well as continued duty of performance'.[20] Apparently, the ILC's view of State responsibility as a negative legal outcome after the breach of a State's obligation by its wrongful act is the same as the concept of liability of the State (*see* §2.03[B]). In the context of space law, the space law community also shares the same opinion as above.[21]

In the context of GNSS and due to its international characteristics (*see* §2.05) determined by its global nature from a technical perspective (*see* §1.02[E]), the terms of responsibility and liability are linked much more to international law. Yet, the terms responsibility and liability have such close ties with the factor of *State* since, currently, most GNSSs are regulated, owned or operated at the same time by the State (for Galileo it is a league of States – the EU, hereafter the same). Even so, in the context of GNSS the demarcation between responsibility and liability should not be vague due to their dual roles in different areas of law.

In private law, the terms responsibility and liability are applied to civil bodies, while, in public law, they are always connected to the State and its representative, i.e., the government. As the provider State combines the status of *civil body* and *public authority* at the same time, both the terms responsibility and liability in the context of GNSS have two levels of connotation as follows:

17. Nathalie Louisa Johanna Theodora Horbach, *Liability Versus Responsibility under International Law: Defending Strict State Responsibility for Transboundary Damage* (N.L.J.T. Horbach, 1996), at 24.
18. *Ibid.*
19. Article 1 of the Responsibility of States for Internationally Wrongful Acts, annex to UN Resolution 56/83.
20. Article 28-31 of the Responsibility of States for Internationally Wrongful Acts, annex to UN Resolution 56/83.
21. Bin Cheng, *International Responsibility and Liability for Launch Activities*, 20 Air & Space Law 1995, at 300; Armel Kerrest, *Remarks on the Responsibility and Liability for Damages Caused by Private Activity in Outer Space*, 40 Proceedings on the Law of Outer Space 1997, at 134.

For the term responsibility:

- the State, as the regulator in public law, is under the duty of supervising the development and operation of GNSS and the provision of GNSS signals/services within its border or jurisdiction, for example, the authorisation for space activities related to GNSS as required by provisions such as Article VI of the Outer Space Treaty;[22]
- the provider State, as GNSS provider, in private law is responsible for ensuring that its provision of GNSS signals/services complies with the required performance parameters (*see* §1.02[D]).[23]

For the term liability:

- the State is liable for its maladministration, which originates from State responsibility for GNSS regulation in public law;
- the provider State is liable for damages caused by the failure of GNSS in private law.

Additionally, when the focus is particularly on space law, the distinction between the terms responsibility and liability is supported by Professor Bin Cheng who defines the former as 'answerability' and the latter as 'obligation to bear the consequences'.[24] The air law community also clearly states that "responsibility under Article 28 should not be seen to be the same as liability," and it does not give private persons a cause of action for compensation in the implementation of CNS/ATM, including GNSS under the Chicago Convention (*see* §4.04[C][3]).[25] Therefore, although the terms responsibility and liability may be misconstrued particularly in general international law,[26] in the context of GNSS they respectively address *prior obligation* and *afterwards effects.*

22. As for which activities are space activities in the development, deployment, and operation of GNSS and the provision of GNSS signal/service, please *see*§3.02[B].
23. In civil aviation sector, as required by Article 15 of the Chicago Convention, each Member State of ICAO should provide air navigation facilities, but whether those facilities compulsorily include GNSS signal/service please *see* §4.04[C][2].
24. The source text is as follows:

> In law, it (responsibility) applies in particular to a person's answerability for compliance with his or her legal duties, and for any breaches thereof.
>
> The term liability is often used specifically to denote the obligation to bear the consequences of a breach of a legal duty, in particular the obligation to make reparation for any damage caused, especially in the form of monetary payment.

Cheng, *supra* note 16.

25. ICAO, *Report on the Establishment of a Legal Framework with Regard to CNS/ATM Systems Including GNSS*, A35-WP/75, LE/5, 28/07/04, at A-4.
26. Frans von der Dunk, *Liability Versus Responsibility in Space Law: Misconception or Misconstruction?*, 34 Proceedings of the International Institute of Space Law 1991, at 363.

[B] Grounds of GNSS Liability: Criminal, Administrative and Civil

Liability, also termed as legal liability, is a basic legal concept with double significance. Aside from the distinction between responsibility and liability as discussed above, the definitions of liability are similar:

> A person is said to be under a liability when he is, or at least may be, legally obliged to do or suffer something. Thus, one may be said to be liable to perform, to pay, to be sued, to be imprisoned, or otherwise to be subject to some legal duty or legal consequence.[27]

In common law systems, liability is composed of *civil* liability and *criminal* liability, which are based on civil law and criminal law respectively.[28] Civil liability is the negative situation after the breach of a compulsory obligation by virtue of, among others, tort law or an agreed obligation subject to contract law, with the legal consequence corresponding *to perform* and/or *to pay* as mentioned above. While civil liability deals with the legal relationship between civil parties with equal status, criminal liability is much more reliant on public power and it is the serious punishment or sanction imposed by a public prosecution organ, in the name of the State, under *nulla poena sine lege (no penalty without a law)* by virtue of criminal law. The legal consequence of criminal liability corresponds to *imprison* or *fine*.

In the context of GNSS, all litigation for civil compensation is without prejudice to parallel criminal proceedings directed against the wrongdoers under national law,[29] where system malfunction or the failure of signal with heavy losses of life and property, and public disorder may trigger criminal liability.[30] For example, China's Criminal Code covers intentional offences such as the 'Crime of disrupting radio communication order' (Article 288), and also non-intentional offences, for example the 'Crime of causing a flight accident' (Article 131), the 'Crime of negligent homicide (Article 233), Crime of major liability accident' (Article 134). The special case of GNSS criminal liability against system owner/operator has at least aroused the EU's awareness.[31]

However, in civil law systems, the liability of administration is treated specially as *administrative liability*, independent of *civil liability* and *criminal liability*.[32] Administrative liability refers to the fact that the State, represented by its government, has

---

27. David M. Walker, *The Oxford Companion to Law* (Clarendon Press, 1980), at 765; Similar definition please *see* Garner, *supra* note 14, at 997.
28. Walker, *Ibid.*
29. EC, *Roadmap: Regulation on EU GNSS third-party liability*, ENTR.GP2, February 2011.
30. Frans von der Dunk, *GNSS Applications – Legal Implications*, presented to Training Course on Global Navigation Satellite Systems and Location Based Services, 4-29 October 2010, African Regional Centre for Space Science and Technology Education in English (ARCSSTE-E), Ile-Ife, Nigeria.
31. European GNSS Programmes Committee, *Global Navigation Satellite Systems (GNSS): Extra Contractual Liability*, EGPC-09-07-06-02, 24 June 2009.
32. The three-way classification of liability is not the so-called *threefold liability rule* in certain jurisdictions according to which *public officials* may be held civilly, criminally, and administratively liable for a violation of their duty or for a wrongful act or omission. However, the term administrative liability in the latter sense is actually an administrative punishment by means of warning, discharge from office or job, monetary penalty, rather than the remedy provided to the

internal sovereignty to its national affairs (*see* §1.04[C]) but should also be responsible for its *faute de service* through paying compensation to the party suffering loss thereof. In other words, administrative liability is the negative outcome after the State or its government fails to fulfil its responsibility on regulation, so much so that it is also called 'State liability' or 'government liability'. Although civil law systems and common law systems treat administrative cases differently in court proceedings, with the former being in a special court (administrative court) and the latter in an ordinary one (civil court), both of them are in fact predominately fault based and do not have large inherent differences in substantive law.[33] Therefore, in the context of GNSS administrative liability there is no need to distinguish purposely the procedural aspects of two legal systems. However, considering that most GNSS providers are currently governments or their agencies, it is necessary to clearly demarcate between 'GNSS administrative liability' and 'GNSS civil liability'. The core nature of the former lies in the maladministration or negligence of supervision and regulation, rather than the factor of whether the liable party is a government or a civil body. Notwithstanding that it is the government, rather than a private company, that would be the provider, GNSS malfunction should be subject to civil law and the civil liability regime.[34]

In all the three grounds of liability, civil liability is addressed much more frequently than criminal liability and administrative liability,[35] and it is the same in space law. In defining the term liability, Professor Bin Cheng also places special emphasis on monetary reparation, which is the main form of civil remedy (*see* above).

---

party who suffered the negative impact by maladministration. *See Dr. Fernando A. Melendres M.D., Executive Director of the Lung Center of the Philippines [LCP] v. President Anti-Graft Commission,* et al., G.R. No. 163859, 15 August 2012; *Office of the President v. Cataquiz,* G.R. No. 183445, 14 September 2011, 657 SCRA 681, 706; *Tecson v. Sandiganbayan,* 376 Phil. 191, 198 (1999) and *Veloso v. Sandiganbayan,* G.R. Nos 89043-65, 16 July 1990, 187 SCRA 504, 509-510; *Second Title: Administrative Liabilities,* Federal Law of Administrative Liabilities of Public Officers, March 13 of 2002, Mexico.

The term *administrative liability* in this book is distinct from *administrative sanction,* which is also usually named as *administrative liability* in some jurisdictions such as China. Administrative sanction is usually imposed on those who fail to comply with the required code of conduct but the event is not serious enough to constitute a crime, or the legal person who does not follow industry standards required by administrative regulations. *See* Danny Pieters & Social Security, *An Introduction to the Basic Principles* (Kluwer Law International, 2006) at 118; Prashant Popat, *International Product Law Manual* (Kluwer Law International, 2010), at 646.

33. *See* Carol Rhian Harlow, *Administrative Liability: A Comparative Study of French and English Law* (University of London, 1979), at 4-6.
34. But this division does not block this book from discussing government liability in the case of GNSS damages caused by a supervisory or a regulatory authority's maladministration or negligence in their supervision duties.
35. Sometimes the term liability is defined directly as civil liability. For example, *Bouvier's Law Dictionary* defines the term liability as follows:

> LIABILITY. Responsibility; the state of one who is bound in law and justice to do something which may be enforced by action. This liability may arise from contracts either express or implied, or in consequence of torts committed.

John Bouvier, *Bouvier's Law Dictionary (Vol. II)* (Philadelpia: George W. Childs, 628 & 630 Chestnut St., 1862), at 41.

In the context of GNSS, it seems that both the legal practice sector and academia seldom make any distinction between 'liability' and 'civil liability', such as the following words in the recommendations made by the ICAO 'Panel of Legal and Technical Experts on the establishment of a Legal Framework with Regard to GNSS (LTEP)' at the very beginning of GNSS-related legal research:

> how liability (emphasis added) provisions concerning the operation, provision and use of GNSS services should ensure that damage arising from such services will be compensated in an equitable manner.[36]

It is obvious that the above description shows no intention and indication connecting the damage to criminal sanction and remedy for maladministration except civil compensation. Although few authors noticed this problem and precisely used the term *civil liability* from the very start,[37] most others just use 'Liability for GNSS' or similar expressions to discuss civil liability for damage caused by GNSS.[38] It is hard to say that the established practice above constitutes a factual misconception or misconstruction of the law related to GNSS, but it is more related to a habit of expression or simplification; conversely, it however demonstrates the core status of civil liability in the framework of GNSS liability in general.

[C] Structure of GNSS Civil Liability: Contractual and Non-contractual

Civil liability is the state of being legally obligated to make good for civil damages, which is imposed by civil law, as opposed to criminal law.[39] Although the term civil liability is deeply involved in international law, including the conventions on air and space law, it is still based on the theory of civil law, even though it is so based from an international perspective with a global scope of application. Civil liability can be either strict/absolute or fault-based; either international, or national, or, in the case of the EU, regional; and either several or joint. In the context of GNSS, civil liability has been a subject in law for a few decades, but few national and international instruments have given a specific definition of GNSS civil liability in both academia and practice. Although ICAO listed certain basic concepts to be considered for further study in

36. ICAO, *Global Air Navigation Plan for CNS/ATM Systems*, Second Edition-2002, Doc 9750, AN/963, at I-11-8.
37. *See*, e.g., Ulrich Magnus, *Civil Liability for Satellite-Based Services*, 13 Uniform Law Review 2008, at 935-969; Pietro Manzini & Anna Masutti, *An International Civil Liability Regime for the Galileo Services: A Proposal*, 33 Air & Space Law 2008, at 114-131.
38. *See*, e.g., Jeffrey A. Rockwell, *Liability of the United States Arising Out of the Civilian Use of the Global Positioning System* (McGill University, 1996), at iv; Gregory Michael, *Legal Issues Including Liability Associated with the Acquisition, Use, and Failure of GPS/GNSS*, 52 The Journal of Navigation 1999, at 246-251; Pablo Rodriguez-Contreras Perez, *GNSS Liability Issues: Possible Solutions to a Global System* (McGill University, 2002), at 1; Frans von der Dunk, *Liability for Global Navigation Satellite Services: A Comparative Analysis of GPS and Galileo*, 30 Journal of Space Law 2004, at 129-167; Hans-Georg Bollweg, *GNSS-Liability by International or European Union Law?*, 59 German Journal of Air and Space Law 2010, at 551-559; Ingo Baumann, *Liability for GNSS Signals and Services*, November/December InsideGNSS 2015, at 38-45; etc.
39. Garner, *supra* note 14, at 997.

relation to the liability regime for GNSS, no mention was made of the concept of civil liability itself.[40] Certain authors may have already noticed this problem, therefore they have introduced the 'concept of liability' as one of the key elements in the proposed 'Liability Regime' of GNSS; however, no further definition is provided.[41] The EU document defines the civil liability in the context of Galileo as follows:

> 'Liability' means the legal accountability of a person or legal entity to compensate for damage caused to another person or legal entity in accordance with specific legal principles and rules. This obligation may be prescribed in an agreement (contractual liability) or in a legal norm (non-contractual liability).[42]

Although the above expression neither makes the liable party clear, nor gives specific considerations to the case of Galileo, it does divide the civil liability into contractual liability and non-contractual liability. In the context of GNSS, contractual liability is based on the privity of contract or agreement and caused by the breach of contractual obligations between contracting parties in the development and deployment of the system, and the provision of PNT signals/services. With regard to the counterpart of contractual liability, this is however being confusingly mingled, to some extent, with the terms of non-contractual liability or extra-contractual liability, tort liability, third-party liability, and product liability. The theory regarding GNSS civil liability structure is now represented by *dichotomy* and *trichotomy*.

The dichotomy expresses that non-contractual liability, i.e., tortious liability,[43] and contractual liability are the *complete* pattern of GNSS civil liability. Furthermore, the former is composed of:

- Non-contractual liability vis-à-vis non-users, that is third-party liability.
- Non-contractual liability vis-à-vis users not bound by contract, such as civil liability claimed by open service users (NCL-UNC).

40. The following concepts, among others, should be considered in relation to the liability regime for GNSS which should be further studied:

    (a) fair, prompt and adequate compensation;
    (b) disclaimer of liability;
    (c) sovereign immunity from jurisdiction;
    (d) physical damage, economic loss, and mental injury;
    (e) joint and several liability;
    (f) recourse action mechanism;
    (g) channelling of liability;
    (h) creation of an international fund (as an additional possibility or an option);
    (i) the two-tier concept, namely strict liability up to a limit to be defined, and fault liability above the ceiling without numerical limits.

    ICAO, *supra* note 36.
41. For example, Bollweg, *supra* note 38.
42. Article 2 of the Cooperation Agreement on a Civil Global Navigation Satellite System (GNSS) – GALILEO between the European Community and its Member States and the People's Republic of China, Beijing, 30 October 2003.
43. Hans-Georg Bollweg, *Initial Considerations Regarding the Feasibility of an International UNIDROIT Instrument to Cover Liability for Damage Caused by Malfunctions in Global (Navigation) Satellite Systems*, 13(4) Uniform Law Review 2008, at 929.

- Non-contractual liability vis-à-vis contractually bound users, such as civil liability claimed by commercial services (NCL-CBU).[44]

In this sense, third-party liability is just one of three subordinate concepts – where product liability is not included – of GNSS civil liability, and the whole pattern is as in Table 2.1.

Table 2.1 Structure of GNSS Civil Liability (Dichotomy-I)

Contractual Liability	Non-contractual Liability		
	Third-party Liability	NCL-UNC	NCL-CBU

Another author followed the dichotomy of civil liability undertaken by GNSS service providers, with contractual liability to contract users and non-contractual liability to third parties. To some extent, this indicates that third-party liability is the *whole* of non-contractual liability in the context of GNSS,[45] as shown in Table 2.2.

Table 2.2 Structure of GNSS Civil Liability (Dichotomy-II)

Contractual Liability	Non-contractual Liability || Third-party Liability

Unlike the above, the trichotomy divides GNSS civil liability into contractual liability, non-contractual liability, and product liability by the 'fundamental threefold distinction' between them.[46] This may be understood to mean that the latter two terms are on an equal level but with a 'fundamental' distinction between them. Meanwhile, some authors see tort liability and third-party liability as the equivalents of non-contractual liability in the context of national legislation and international law respectively.[47] This structure is outlined almost in the same way by yet again another author, only with one more reason, i.e., the different conflict-of-law rules applicable to the above three terms. Based on their words, the picture of GNSS civil liability-related terms could be illustrated in Table 2.3.

Table 2.3 Structure of GNSS Civil Liability (Trichotomy)

Contractual Liability	Non-contractual Liability || Tort Liability || Third-party Liability	Product Liability

44. Bollweg, *supra* note 38.
45. Jingjing Nie, *The Future of Uniform Rules on GNSS Liability*, 54 Proceedings of the International Institute of Space Law 2011, at 339-340.
46. *See* von der Dunk, *supra* note 38.
47. *Ibid.*

Although no heated argument has arisen about the *dichotomy* and *trichotomy* of GNSS civil liability and their internal lack of conformity, the disputes related to these are still obvious in the three fairly simplistic tables above, which can be roughly summarised in the form of the following questions:

(i) What is the complementary concept of GNSS contractual liability?

As both 'extra-' and 'non-' are used as negative prefixes in English, which mean 'beyond', 'outside' or 'not', it is safe to say that non-contractual liability and extra-contractual liability are appropriate counter partners, or 'antonyms' in a more informal way, of contractual liability with the same essence but different outward forms. The main divergence concerns the subordination between non-contractual liability and, in particular, tort liability or liability in tort, which is quite complicated, or, let us say, confused, not only in the context of GNSS but also in a broader sense.

There is one opinion both in theory,[48] and in practice, i.e., legislation,[49] which views extra-contractual liability and tort liability as just two ways of expressing ideas in civil law and common law respectively.[50] In addition, the latest revised 'Code civil des Français' uses the term 'la responsabilité extracontractuelle' (extra-contractual liability),[51] instead of its predecessor 'des délits et des quasi-délits' (intentional and unintentional wrongs [Of Torts]),[52] and this provides indirect evidence of the fact that, now, French legislation does not distinguish tort liability from non-contractual liability.

However, from the European perspective as a whole, non-contractual liability seems to be a much broader concept than tort liability. The 'Regulation on the law applicable to non-contractual obligations' (Rome II Regulation)' states, in defining the term 'non-contractual liability',[53] that "damages shall cover any consequence arising out of tort/delict, unjust enrichment, *negotiorum gestio* or *culpa in contrahendo,*" which corresponds with non-contractual liability arising out of tort/delict, i.e., tort liability, and non-contractual liability arising out of unjust enrichment,[54] *negotiorum gestio* or *culpa in contrahendo.* Consistency with Rome II Regulation is kept to some extent in the draft of the European Civil Code, according to which, the law on

48. For example, Daniel Rubiano Rincon, *Environmental Law in Colombia* (Kluwer Law International, 2011), at 149.
49. For example, Federal Law-Civil Law Harmonization Act of Canada, No. 1, S.C. 2001, c. 4, s. 34.
50. *See Bijural Terminology Records*, http://justice.gc.ca/eng/csj-sjc/harmonization/bijurilex/terminolog/not176.html, last accessed 2 March 2015.
51. Ordonnance n° 2016-131 du 10 février 2016 portant réforme du droit des contrats, du régime général et de la preuve des obligations, NOR: JUSC1522466R, https://www.legifrance.gouv.fr/eli/ordonnance/2016/2/10/JUSC1522466R/jo/texte, last accessed 6 May 2016.
52. Chapitre II, Titre IV, Livre III of the Code civil des Français and its English version, https://www.legifrance.gouv.fr/content/download/1950/13681/version/3/file/Code_22.pdf, last accessed 6 May 2016.
53. Article 2 of the Regulation (EC) No 864/2007 of the European Parliament and of the Council of 11 July 2007 on the law applicable to non-contractual obligations.
54. The researcher in common law system calls civil liability for unjust enrichment as matters of restitution which is one of related bodies of law together with tort and contract. *See* Ward Farnsworth, *Restitution: Civil Liability for Unjust Enrichment* (University of Chicago Press, 2014), at 1.

non-contractual liability is composed of the law on non-contractual liability for damage (Book VI), the law on unjustified enrichment (Book VII), and the rules on benevolent intervention in another's affairs (Book V).[55] The law of non-contractual liability arising out of damage caused to another, in contrast to other types of non-contractual liability, could be called tort law or the law of tort,[56] even though that is not the whole of non-contractual liability law.[57] Furthermore, the above approach could also find supporting points from the *Bürgerliches Gesetzbuch*[58] and the *Burgerlijk Wetboek*.[59]

Here is not the place to make an evaluation of the merits and demerits of the above two approaches, however technically speaking, the European way is much more logical, particularly when considering that it is backed up by the well-known German precise characteristic, in contrast with French romanticism. Further, taking into account the much broader representative of the *European* way and considering the usage of 'extra-' or 'non-' in the English language, it could be hence concluded that the relationship between non-contractual/extra-contractual liability and tort liability is superior-subordinate rather than parallel; in other words, the former is one of the branches of the latter. Nevertheless, unlike the traditional case of private law, GNSS or its owner/operator shows the connection with unjust enrichment, *negotiorum gestio* or *culpa in contrahendo* next to nothing in the damages caused by GNSS malfunction or failure. Therefore, in the context of GNSS, tort liability indeed constitutes most, but still not the whole picture of non-contractual liability.

(ii) What is the relationship between GNSS third-party liability and GNSS tort liability?

The term third-party liability seems to be recognised more by international organisations and conventions. This is particularly so in the legal regime of nuclear energy[60] and aviation transportation,[61] both of which activities are labelled *ultra-hazardous*.[62]

55. Christian von Bar, et al. (Eds), *Principles, Definitions and Model Rules of European Private Law Draft: Common Frame of Reference (DCFR)-Outline Edition* (Sellier. European Law Publishers, 2009), at 80.
56. Christian von Bar, *Non-Contractual Liability Arising out of Damage Caused to Another* (sellier. european law publishers, 2009), at 229.
57. Christian von Bar & Ulrich Drobnig, *The Interaction of Contract Law and Tort and Property Law in Europe: A Comparative Study* (sellier. european law publishers, 2004), at 307.
58. Division 3 of Book 2 of the *Bürgerliches Gesetzbuch* is entitled *Schuldverhältnisse aus Verträgen* (Contractual Obligations), and title 27 *Schuldverhältnisse aus Verträgen* (Torts) and title 26 *Ungerechtfertigte Bereicherung* (Unjust Enrichment) are categorised under Division 8 *Einzelne Schuldverhältnisse* (Particular Types of Obligations) thereof.
59. Title 4 under Book 6 of the *Burgerlijk Wetboek* is named as *Verbintenissen uit andere bron dan onrechtmatige daad of overeenkomst* (obligations from another legal source than tort or contract) immediately after title 3 *Onrechtmatige daad* (Tort).
60. For example, the Convention on Third-party liability in the Field of Nuclear Energy of 29 July 1960, as amended by the Additional Protocol of 28 January 1964 and by the Protocol of 16 November 1982.
61. For example, the Rome Convention on Damage Caused by Foreign Aircraft to Third Parties on the Surface, on 7 October 1952.
62. Some authors point out that aviation was considered as an ultra-hazardous activity, but it is no longer applied in all aviation cases instead some specific types of aviation types such as testing experimental aircraft. *See* Michael W Pearson & Daniel S. Riley, *Foundations of Aviation Law* (Ashgate Publishing, Ltd, 2015), at 52.

Even though it is true that in the text of national legislation the term third-party liability rarely appears instead of the term tort liability, in the context of GNSS it is neither enough, nor appropriate to say that the two terms have the same meaning.

The author is not going to deny the fact that contributors beyond the national level to GNSS law or policy, such as the International Institute for the Unification of Private Law (UNIDROIT), the EU and the European Space Policy Institute (ESPI),[63] favour the expression 'third-party liability for GNSS' or 'GNSS third-party liability' much more than they do 'GNSS tort liability'. However, the difference between the term third-party liability and tort liability concerns not only the respective mode of expression in international law or national legislation, and in civil law or common law, but also the definition itself.

The core of the term 'third-party liability', also known as 'tort liability to/for third parties', is 'third party', which refers to "any person who is not a party to a relationship or transaction between any two others."[64] In these words, 'third party' is based on the other two parties having a relationship, which is usually, but not limited to a contractual relationship. In terms of liability arrangement, there is also no contractual relationship under the name of a third party.[65] This directs us to the simple understanding that, when there are no first and second parties there is no third party, and therefore there is no third-party liability. In the context of GNSS, the first and second parties are GNSS providers, either basic signal providers or augmented signal providers, and the user; the third party is the group of victims in the accident caused by GNSS, which can be illustrated by the two following examples with regard to EU GNSS included in the EU document:

(i) with regard to the Open Service: a car accident, where the driver is the user of the GNSS, and the pedestrians walking by are the third parties;
(ii) with regard to the Safety-of-life Service: two aircraft colliding over a populated area, where the user is the aviation company, and the third parties are the passengers on board – though covered by a particular international Convention – and the victims on the ground.[66]

Combining the definition of third-party liability in general, that is the civil liability towards a third party, we can read from the above examples that the GNSS third-party liability regime does not apply to the damage caused by GNSS providers to the users. This is similar to the case where the third-party liability insurance does not cover the damage to the insurance contractors at issue.

Does this however mean that the civil disputes between the two parties other than the third – constituting, in the two examples above, civil liability of GNSS providers to the driver and aviation company for their loss of property and bodily

63. *See* UNIDROIT, *An Instrument on Third Party Liability for Global Navigation Satellite System (GNSS) Services: A Preliminary Study* (UNIDROIT, 2010), Study LXXIX – Preliminary Study, March 2010; EC, *supra* note 29; ESPI, *Policy Aspects of Third-Party Liability in Satellite Navigation: Preparing a Roadmap for Europe*, Report 19, P42-C20490-04, July 2009.
64. Walker, *supra* note 27, at 1216.
65. Joseph N. Pelton & Ram Jakhu, *Space Safety Regulations and Standards* (Elsevier, 2010), at 208.
66. EC, *supra* note 29.

injuries of employees – have to resort to the GNSS contractual liability regime? The answer is clearly in the negative for the following three reasons:

First, the question whether there is an 'implied contract' between GNSS open service/signal providers and relevant users remains open (*see* §5.02[C][2]). If the contractual relationship between the above two parties is not recognised, which is the most common case, the open signal user/the driver that suffered damage has to claim for compensation based on *general tort law*, rather than the law on third-party liability.

Second, to say the least, given that the contractual relationship is accepted in open service by the court, or is established in the commercial service, under the regime of *anspruchskonkurrenz* in German law[67] as well as Chinese law[68] or the concept of *contort*, i.e., contract plus tort, in English law,[69] the user that suffered damage is allowed to claim for compensation against the GNSS provider based on *general tort liability*, not third-party liability, even though that user is entitled to claim contractual liability at the same time.[70]

Third, unlike the term GNSS product liability, which usually applies strict liability (*see* below), GNSS third-party liability does not have its own featured regime and has to depend on the general theory of tort liability, even though legislators or researchers use an independent term simply because it is the collective concept of civil liability to/for a third party. Therefore, it can be concluded that, regardless of how the term is used in international law or national legislation, third-party liability is just a specific category of tort liability,[71] which is supplemented by general tort liability in the context of GNSS. The term GNSS third-party liability appears only in those cases where there are second parties, these being the GNSS providers and the users, and therefore there are third parties, these being the victims suffering damage caused by GNSS.

(iii) Is the term product liability unique enough to be qualified as one of independent pillars of GNSS civil liability?

Although unpopular, some academic researchers,[72] rather than rule-making bodies, do address GNSS civil liability from the perspective of product liability. The author is not

67. Apostolos Georgiades, Die Anspruchskonkurrenz im Zivilrecht und Zivilprozeßrecht (Beck, 1968), at 167.
68. Article 122 of China's Contract Law reads as follows:

> Where the breach of contract by one party infringes upon the other party's personal or property rights, the aggrieved party is entitled to choose to claim the assumption by the violating and infringing party of liabilities for breach of contract according to this Law, or to claim the assumption by the violating and infringing party of liabilities for infringement according to other laws.

69. *See* Garner, *supra* note 14, at 365; *Brown -v- Boorman, Boorman, and Wild*, [1844] EngR 65, (1844) 11 Cl & Fin 1, (1844) 8 ER 1003.
70. If some torts are also breach of contract, the party suffering damage may sue either in tort or for breach of contract, or both. *See* Jonathan Law, *A Dictionary of Law* (Oxford University Press, 2015), at 623.
71. *See* Claire McIvor, *Third-Party Liability in Tort* (Bloomsbury Publishing, 2006), at 1.
72. *See*, e.g., Andreas Loukakis, *Product Liability Ramifications for Erroneous GNSS Signals: An Alternative Approach Is Possible?*, 56 Proceedings of the International Institute of Space Law

going to discuss how and to what extent product liability can be applied to GNSS signals/services or whether it is necessary to create a specific system named *GNSS product liability*; what the author is going to do is determine the position of the term product liability in the whole definition framework of GNSS civil liability.

Compared with legal concepts such as contract and negligence, product liability is defined more empirically as the liability of manufacturer and other persons for defective products.[73] Regardless of 'non-strict' theories of product liability in US law, where negligence, tortious misrepresentation and breach of warranty may be used as ground for a claim,[74] the term product liability is indeed more closely connected to *strict liability*, which is different from the general theory of tort liability based on fault. Considering the special rules of product liability, besides academic research such as the 'American Restatement (Third) of Tort Law: Products Liability', many national laws codify this division of civil liability independently, for example, Title IV bis 'Of Liability for Defective Product' of the French Civil Code. In addition, many states in the US have enacted comprehensive product liability statutes, and a Model Uniform Products Liability Act was promulgated to make up for the lack of Federal products liability law.[75] More convincingly, a specific directive on defective product liability has been effective since as early as 1985.[76]

In the context of GNSS, the distinctiveness of product liability seems more obvious considering two possible ways of understanding the term GNSS product liability under the regime of GNSS civil liability:

(i) a GNSS satellite or its components, user equipment having a defect, and the defective software, system design, or workflow in the circulation of GNSS operation may trigger product liability in a more usual way if damage is thus caused;
(ii) whether a GNSS signal with wrong PNT data or information which lead to damage may be considered as a defective product and then apply the product liability regime, with the reference to such analogies as electricity or defective aeronautical charts.[77]

More importantly, GNSS product liability may direct victims to seek compensation from the manufacturer or relevant civilian bodies bearing the responsibility for guaranteeing safety in the circulation of products, and this is without the difficulty, in most cases, of having to prove fault thereof under the umbrella of strict liability. This makes the liable party far more identifiable than under the general tort liability regime.

2013, at 320-324; Frans von der Dunk, *GNSS Applications-Legal implications*, presented to UN Office for Outer Space Affairs, 6 October 2010, at 61-67; Qi Gao, *Civil Liability of GNSS Service Provider: From the Perspective of American Law and Practice*, 29(2) Journal of Beijing University of Aeronautics and Astronautics (Social Science Edition) 2016, in Chinese, at 30-31.

73. Jonathan Law, *supra* note 70, at 484.
74. *See* Michael Krauss, *Principles of Products Liability* (West Academic, 2014), at 55.
75. *Products liability*, Legal Information Institute, Cornell Universality Law School, https://www.law.cornell.edu/wex/products_liability, last accessed 15 May 2016.
76. Council Directive 85/374/EEC of 25 July 1985 on the approximation of the laws, regulations and administrative provisions of the Member States concerning liability for defective products.
77. Loukakis, *supra* note 72.

In addition, next to the term GNSS product liability, *service liability*, which is usually fault-based liability, is viewed by at least one author as a more suitable concept for GNSS;[78] at the current time this question is indeed confusing since neither has the legal nature of space signals been settled, nor has it been settled whether GNSS provides only signals, whether it could be qualified as one kind of *service*, or whether it varies on a case-by-case basis.

Based on the distinctiveness of product liability and its *value* to victims as well as complexity in the context of GNSS, it will be nice to characterise GNSS product liability as an independent pillar of GNSS civil liability besides GNSS contractual liability and GNSS tort liability. The following clarifications have to be however clear on the subordination of respective terms in the definition of GNSS civil liability:

First, GNSS product liability falls under the category of GNSS tort liability as a subset. Although the application of product liability ramification does not exclude the parties with a contractual relationship, and regardless of how GNSS product liability is interpreted, the term is still a concept under the framework of tort liability since its legal basis is not on the contract in question but the law, including both case law, and statutory law namely tort law. The law requires that providers, including the system constructors, equipment manufacturers, operators, and other actors involved in the circulation of GNSS products, guarantee that what they offer to the consumer or someone to whom the product was transferred is of such a level of safety as persons are generally entitled to expect. Otherwise, if not, they are obligated to make fair remedy to the party suffering damage. The difference between tort liability and product liability in the context of GNSS is not *fundamental* but *subordinate*. In addition, even the researcher who addresses the significance of the *threefold distinction between tort, contractual and product liability* has to respect the fact that *product liability can be defined as a specific type of tort liability* in the context of GNSS.[79]

Second, the scopes of *GNSS product liability* and *GNSS third-party liability* sometimes overlap and hence they are not in a relationship of *either A or non-A* under the framework of GNSS tort liability. The definition of product liability focuses only on whether the liability is caused by a defective product and does not consider whether or not victims are a third party. For example, in an autopilot aircraft accident caused by the malfunction of the GLS due to incorrect PNT signals, the passengers, as third parties who are outside the transaction of the PNT signals, could also claim compensation against the signal *producer* in question, if GNSS signals are recognised as the subject of product liability law.[80]

Based on the analysis above, it is clear that contractual liability and non-contractual liability constitute the complete picture of civil liability in the context of GNSS. Although it is not the entire picture, GNSS tort liability represents the major percentage of GNSS non-contractual liability. Both GNSS third-party liability and GNSS product liability are two subordinate concepts under the same 'guiding' GNSS tort

78. Atsuyo Ito, *Legal Aspects of Satellite Remote Sensing* (Martinus Nijhoff Publishers, 2011), at 282.
79. Loukakis, *supra* note 72.
80. In this case, the question that whether the GLS constructor, the airport or the GNSS provider, or all of them, are the *producer* of incorrect PNT signals remains open at the current stage.

liability, and although not complementary, as different criteria of classification they do occasionally overlap. The pattern above could be illustrated in Table 2.4.

Table 2.4 Structure of GNSS Civil Liability (Author)

<table>
<tr><td rowspan="2">Contractual Liability</td><td colspan="2">Non-contractual Liability</td></tr>
<tr><td>≈
Tort Liability
Third-party Liability
and/or
Product Liability
and/or
other types of Tort Liability</td><td>...</td></tr>
</table>

The three concepts in italic in Table 2.4 constitute three pillars of the GNSS civil liability regime, namely:

(i) *GNSS contractual liability*, triggered by the breach of contract of the provision of GNSS signals or services.
(ii) *GNSS tort liability (without product liability)*, hereinafter referred as *GNSS general tort liability*, triggered by an unlawful act of the GNSS provider, excluding the defects of the GNSS product defined below.
(iii) *GNSS product liability*, triggered by a defective GNSS product, including space satellite, ground stations, and their components for the provision of GNSS signals or services, and GNSS signals themselves but whose qualification as a product remains to be discussed further.

§2.04 ELEMENTS IN ESTABLISHING GNSS CIVIL LIABILITY

[A] Four Elements of GNSS Civil Liability

The discussion above hopefully clarifies the respective relationships of different concepts under the framework of GNSS civil liability. Nevertheless, in order to establish GNSS civil liability and be able to benefit from it in court, certain elements must be proven. Although the content of those elements depends on the national legislation applied to the case of GNSS damage and on the basis of civil liability relied on by the claimant, it is still possible to extract the basic formula. Compared to the history of legal theory on civil liability from as far back as Roman law, GNSS civil liability is obviously a new area, meaning that it has to be rooted in a general theory of civil liability. By recalling the definition of civil liability and referring to case law, regardless of the nature of the claim in a civil case, two parties – claimant and defendant – must be qualified, and two facts must be established: the defendant did something which violated either legal or contractual rules, and the health and/or economic interests of the claimant are *therefore* damaged or under significant threat. This also holds true for GNSS civil liability. Hence, the following sections discusses the elements, in the context of GNSS, of: *the parties*; *unreasonable acts*; *damage*; and their *causality*.

[B] The Parties: GNSS Providers Versus GNSS Users or/and Third Parties

Without considering the complexity of the GNSS technical constitution and legal relationships, the roles of the parties in court are usually obvious: the victim suffering the damage caused by GNSS and/or its family acting as the potential claimant, and the potentially liable party providing GNSS signals/services and, if any, its insurer summoned to appear in court as the defendant.[81] Figure 1.3 above shows how intricate the various lines, representative of relationships, between all actors are, and this means that we need to reasonably *translate* those technical participants into legal subjects, especially the liable party (compensator) and potential victims (claimer). Throughout the entire process of making PNT signals finally meet up with the user, two technical groups represent those two parties in the supply and demand chain of GNSS respectively, these being *GNSS providers* versus *GNSS users*. Outside that chain, a *third party* could also be involved as either compensator or claimer depending on an individual case.

(i) GNSS providers.

Aside from the *direct provider* to transmit PNT signals to the users, there are various actors from the supply side in the value chain of GNSS, including but not limited to the *owner, system constructor,* and *operator*. A GNSS provider is the party most probable to be the main defendant in court. Under the current pattern of the GNSS industry, most GNSS providers are the *State* or a public authority in nature, that is, the regulator and legislator at the same time. This gives rise to the applicability of the principle of State immunity (*see* §5.03[B][2]).[82] In addition, as few States can afford their own GNSS, they have to subcontract the responsibility of GNSS provision to foreigners.

Although some actors from the supply side in the value chain of GNSS are the same single body, it is not always the case.[83] The operations of space satellites and

81. The insurer can also be the direct defendant, but only as the party who pays compensation to the victims, rather than the liable party. *See* Sergio M. Carbone & Maria Elena De Maestri, *The Rationale for an International Convention on Third-Party Liability for Satellite Navigation Signals*, 14(1-2) Uniform Law Review 2009, at 54.
82. Here the big difference between 'public authority' and 'public company' should be noted: the former enjoys immunity under the umbrella of State immunity, while the latter has the same liability and tax issues as a private company. EC, *Proposal for a Regulation of the European Parliament and of the Council on Further Implementation of the European Satellite Navigation Programmes (2014-2020)*, SEC (2011) 1446 final, Brussels, 30 November 2011.
83. Regardless, the owner, constructor, operator, and direct provider of GNSS signals/services are not always the same single public authority. Take one basic GNSS system and one augmentation system respectively as examples:

> For the Galileo system as a GNSS, in the In-Orbit Validation phase, the ESA is the developer and system designer, European Commission (EC) is the co-funder; In the Full

ground facilities may be granted to different bodies, while GNSS providers can be more than one entity, all of which focus on specific areas such as aviation or maritime, and/or specific types of GNSS activity respectively. Currently, most actors from the supply side in the value chain of GNSS are public in nature, while some private contractors or suppliers may be directly involved into court proceedings particularly in the case of product liability. Furthermore, if a GNSS signal is recognised as 'product' in court proceedings, those parties in each link of the chain would be held liable as either manufacturer or seller.[84]

The plurality of possible defendants offers more sources of compensation but decreases the *identifiability* of the liable party at the same time. Therefore, *channelling of liability* and *joint liability* are much needed in the context of GNSS to guarantee, to the claimant, the easy identifiability of the liable party. Also, that plurality increases the possibility of *vicarious* liability, and this will lead to certain recourse actions where the defendant in one case could be the claimant in another case against the actual liable party.

(ii) GNSS users.

According to what a GNSS provider provides, a GNSS user could be categorised as either an 'open service user' or an 'authorised service user'. While the eligibility of the latter as a possible claimant is usually recognised on the basis of a contractual relationship, in the case of the former it is difficult to find the legal basis. It is self-evident that open service users do not establish the required contractual relationship with a GNSS provider, making it less possible to submit a claim on breach of contract (*see* §5.02[C][2]). Furthermore, it is very hard to establish a high-level duty of care to the public at large in a highly attenuated relationship[85] as the potential number of persons to whom an open service might become available is without limit (*see* §5.02[C][3]).

In addition, the term GNSS users is a multilevel concept. In other words, before a GNSS signal is finally transmitted to its terminal user, each provider, compared with its downstream user, is also the signal user of its upstream provider and they can be

Operational Capability phase, the funder and owner would be the EC alone, and the ESA is the only agent of the EC for the role of design and procurement. *See The History of Galileo*, http://ec.europa.eu/growth/sectors/space/galileo/history/index_en.htm, last accessed 16 May 2016.

The GAGAN augmentation satellite, an India augmentation satellite system, is a joint venture enterprise between Indian Space Research Organization (ISRO) and the AAI. While ISRO operates the satellite and the command and telemetry control, indeed as an operator, it is the AAI that will be the direct provider of the GPS/GAGAN signals for air navigation services. *See* Ranjana Kaul, *India-Liability in Context to the Air Navigation Service Provider*, presented to the International Conference on Contemporary Issues in Air Transport, Air Law & Regulation, New Delhi, India, 23-25 April 2008.

84. Product liability is not linked to the direct manufacturer of components, and if that component is installed into another product, the producer of the whole product could also be sued through product liability. The case should be the same in the context of GNSS. *See MacPherson v. Buick Motor Co.*, 160 A.D. 55, 145 N.Y.S. 462, N.Y.A.D. 3 Dept. 1914.
85. *Rosenberg v. Harwood et al*, Utah District Court, Case No. 2:10-cv-00496.

called as 'first user', 'second user' and so on all the way down to the 'final user'. The final point that needs to be addressed is that the international characteristics of GNSS users inevitably make a big difference in identifying the claimant and jurisdiction in the cases of GNSS civil liability (*see* §2.05).

(iii) Third parties.

As mentioned, a third party is beyond any relationship or transaction between the other two parties. As the victim who suffered GNSS damage, a third party could be the subject of third-party liability; yet, as the source of frequency interference against GNSS signals, a third party could potentially be the liable party, regardless of intent. Considering the pervasiveness and unpredictability of a third party, it would be a good solution to channel its civil liability to a more identifiable party such as GNSS providers.

In addition, unlike a GNSS provider, which is usually a public authority or sizeable corporation, a third party could be a five-year-old child or mental health patient who, in any case, would be the victim. To this end, if such a third party were the wrongdoer, the guardian thereof would have to assume the 'vicarious liability of guardians', this being a type of tort liability.[86]

[C] Triggers: Unreasonable Acts

GNSSs are facing technical vulnerability, financial pressure and institutional challenges (*see* §1.03), but can all those elements be transferred to the triggers of civil liability? Not really. Two typical cases are:

(i) *State of the art technology*. No matter how advanced GNSS technology is declared to be, it will always have a gap between real and ideological condition. A technical defect which is beyond current practical technological feasibility is free of civil liability, even under the regime of strict liability, namely, product liability.[87]

(ii) *Acts of God*. GNSS signals pass from the near vacuum of space, and then through the various layers of the atmosphere to the earth at the speed of light.[88] In doing so, it is unlikely that it will avoid being influenced by uncontrolled space weather such as ionosphere and solar activity which would lead to inaccurate or total loss of PNT signals, and therefore cause damages.[89] Similar situations may also occur because of earthquakes, tsunamis, and other natural disasters which may destroy the ground facilities of

86. Jaap Muscle & Francesco Donato Busnelli, *Unification of Tort Law: Liability for Damage Caused by Others* (Kluwer Law International, 2003), at 239.
87. For example, Article 7 (e), Council Directive 85/374/EEC; § 107 (A) (D), the US Model Uniform Product Liability Act; Article 1386-11, French Civil Code.
88. NovAtel Inc., *An Introduction to GNSS* (NovAtel Inc., 2010), at 20.
89. The errors caused by bad space weather thereof maybe very small, but keep in mind that in one nanosecond, light travels thirty centimeters, and this level of inaccuracy is enough to cause disasters. *Ibid.*

GNSS.[90] In the legal sense, no one body may be liable for events outside human control.[91]

The conditions above free the parties from civil liability regardless of contract law or tort law, civil law systems or common law systems. The legal causes of GNSS civil liability could be divided into the following three branches, which correspond to the three pillars of GNSS civil liability (*see* §2.03[C]), because different rules will apply and diverse subjective requirements follow from this distinction.

(i) *Wrongful act or omission*. Generally speaking, civil liability will arise only upon the proven existence of the fault of the obligated person, which may be expressed by act or omission. In the context of GNSS, civil liability is triggered by the failure of GNSS signals, including the absence of signals, the error of signals, and/or the degradation of performance.[92] The two main causes giving rise to the failure of GNSS signals can be summarised as follows:

(a) *system malfunction*, which could be caused by a navigation satellite crash just like the 2009 Iridium-Cosmos Collision, system shutdown or, more possibly, partial signal cut-off,[93] incorrect uploading of data,[94] organisational problems, and other malfunctions which can be attributed to the provider's intention or negligence;

(b) *harmful interference*, which refers to the radio frequency conflicts from jammers and spoofing devices, in particular the low-cost low-priced ones, which are usually wilfully or unintentionally used by third parties including privacy rights fighters, government authorised services providers, operating forces, criminals, or, even worse, terrorists, etc.[95]

---

90. The distinction between 'acts of God' and 'force majeure' lies in: the former is limited to a natural disaster which is beyond human control; in addition to natural disasters, the latter also includes societal events impossible to be controlled or anticipated by a single party, including acts of war, acts of terrorism, etc. *See* Marsha L. Baum, *When Nature Strikes: Weather Disasters and the Law* (Greenwood Publishing Group, 2007), at 87.

91. As GNSS connects with acts of war and acts of terrorist much more than other non-safety-of-life system, whether the war or terrorist event could be cited as a defence in particular by a State, remains open.

92. Anna Masutti, *CSN/ATM Systems: Framework and Regulation on GNSS, Experiences in Europe*, presented to the Conference on Contemporary Issues in Air Transport, Air Law and Regulation, New Delhi, India, 21-25 April 2008.

93. This may happen during military or political conflicts and this is the exact reason why China developed its own BDS.

94. *See Satellite Outages Afflict GLONASS: The Russian System Suffers Two Major Disruptions in April*, http://www.insidegnss.com/node/4009, last accessed 24 May 2016.

95. The threat of jamming and spoofing technology is not unreachable any more to the safety of common people's lives. For example, in August 2012, a US citizen operated unlawfully a USD 33 GPS jamming device that caused harmful interference to the GBAS used for precision approach, departure procedures, and terminal area operations in Newark Liberty International Airport (Federal Communications Commission (FCC), *Notice of Apparent Liability for Forfeiture*, FCC 13-106, 1 August 2013); in January 2007, as two navy ships were conducting test procedures when communications were lost by jamming radio signals, air traffic management systems, emergency systems, and cell phone communication, and ATMs in San Diego, California were unwittingly disabled for about two hours; in 2013, an academic group from the University of

In strict liability, where a victim suffers damage without anyone deserving blame for it,[96] the element of wrongful act or omission is still however required, otherwise it would be the case of 'acts of God' (*see* above). Nevertheless, if the subject of that wrongful act or omission cannot be attributed to the defendant, the actual claim against the defendant is assumed to be vicarious liability. For example, when certain civilian GNSS providers have no choice but to implement national policy, like the former SA policy did so secretly and suddenly for the sake of 'national interest', the behaviour of that government constitutes a wrongful act even though the civilian GNSS provider has to bear civil liability if damage happens because of that.[97]

(ii) *Non-performance*. This term here is limited to the sense of the failure to perform an agreed obligation in a reasonable manner, for example, breach of contract. To judge non-performance in the context of GNSS, the court or expert witness would need to proceed from the four performance parameters: accuracy, integrity, continuity, and availability (*see* §1.02[D]). The detailed indicators or the general description of level of performance required is found in the contract terms or minimum standards by law. In general, if a GNSS provider itself fails to comply with required and promised performance parameters, the provider would be considered to undertake the contractual liability.

As to the requirement of subjective matter, the most striking rule of civil liability in contract law is no fault, compared with the fault rule in tort law, even though the special tort liability regime, such as product liability, blurs the line of demarcation.[98] Although certain authors researching for the 'Economics of Law' are trying to address fault in contract law,[99] national

Texas demonstrated how a false GPS signal generator could override a luxury yacht's navigation computers as it travelled at sea; a study in 2011 on the problem at Taiwan's Kaohsiung International Airport found an average of 177 jamming and spoofing incidents a day; in 2011, Iran successfully hijacked the US UAV through GPS spoofing; etc. All the above news can be publicly accessed from the Internet.

96. Jaap Hage, et al. (Eds), *Introduction to Law* (Springer International Publishing Switzerland, 2014), at 104.
97. The cut-throat competition may happen in the context of GNSS between space rivals. For example, Roscosmos (Russian Federal Space Agency, which has been transferred to the 'State Corporation for Space Activities') was threatening to restrict the use of certain GPS ground stations located in Russian territory because its attempts to build a base in the US were blocked by the US government. *See* Jeo Miller, *Russia to 'Restrict' US-Run GPS Satellites*, http://www.bbc.com/news/technology-27662580, last accessed 22 September 2015; ***Об исполнении поручения Правительства Российской Федерации***, http://www.federalspace.ru/20646/, last accessed 22 September 2015.
98. Omri Ben-Shahar & Ariel Porat, *Fault in American Contract Law*, 107 Michigan Law Review 2009, at 1344.
99. *See*, e.g., Melvin Aron Eisenberg, *The Role of Fault in Contract Law: Unconscionability, Unexpected Circumstances, Interpretation, Mistake, and Nonperformance*, 107 Michigan Law Review 2009, at 1413; Richard A. Epstein, *The Many Faces of Fault in Contract Law: Or How to Do Economics Right, Without Really Trying*, 107 Michigan Law Review 2009, at 1461; George M. Cohen, *The Fault That Lies Within Our Contract Law*, 107 Michigan Law Review 2009, at 1445.

legislation rarely considers that factor in the case of contract liability.[100] In addition, bearing in mind the technical complexity of GNSS, the stricter the obligation, the easier it is for the promisee who complains of a breach to establish liability.[101] So far there is no revolutionary excuse to establish fault in GNSS contractual liability, which would be against the general theory of contract law. Hence fault, regardless of wilful or negligent breach of contract, is not a necessary element to establish GNSS contractual liability.

(iii) *Defects in GNSS provision.* As one form of typical strict liability, product liability seems never to take into account whether it is the manufacturer's or the seller's fault. The victim may claim for no-fault liability for the damage caused by the defective GNSS equipment or signals/services (*see* §2.03[C]). However, whether that which a GNSS provides can be labelled as a 'service', 'product', 'data', 'information' or just 'signal' remains open.

[D] Damage: Safety-of-Life Dangers

Satellite navigation, satellite communication, and satellite remote sensing are three critical applications of satellites with the same importance. Why, then, is civil liability a topical concern of satellite navigation, but seldom an issue in either satellite communication or satellite remote sensing? The reason behind this phenomenon is the gaping differences on the extent of damage caused by each of them. Unlike satellite communication, the service based on GNSS is related much more to safety of life, especially in the transportation and SAR sectors, even though this can also connect terrorists by easy spoofing devices. Moreover, safety-critical applications of GNSS usually need more real-time data than satellite communication, especially for high-speed mobile vehicles such as landing aircraft. Likewise, satellite remote sensing is actually a space-based monitoring system which is not capable of mitigating ongoing disasters. The reason why GNSS has more to do with dangers lies in the fact that it can play the role of the controller of autopiloted aircraft and other decision-making systems, whereas satellite communication and remote sensing are usually the assistant tools of the controller's decision.

As to the types of damage caused by GNSS, they range from inconvenience to catastrophe:[102]

- *personal injury*, which is always the key issue even in the context of GNSS; yet, whether mental injury is included should be decided by the national law selected by the court;

100. For example, China's Contract Law regulates contractual liability as one strict liability, i.e., no-fault liability. *See* Guangxing Zhang & Shiyuan Han, *General Principles of Contract Law* (Law Press China, 1999), in Chinese, at 86.
101. Solène Rowan, *Fault and Breach of Contract in France and England: Some Comparisons*, 22(4) European Business Law Review 2011, at 467.
102. John A. Lever, *Unintended Consequences of the Global Positioning System*, 7 Systems Engineering 2004, at 221.

- *economic loss*, including the property damage and the delay of goods or passengers because of the GNSS malfunctioning, in particular in the civil aviation sector;
- *environmental damage*, which may happen, for example, if the GNSS causes the collision of vehicles transporting oil, nuclear or other dangerous materials, or damage to the nuclear vehicle itself;[103]
- *privacy leak*, caused by an unlawful GNSS data leak from a provider or because of system-related reasons, such as suffering a hacker attack, rather than the illegal use of GNSS equipment (*see* (v) of §1.04[C]);
- *other* types of detrimental effect.

This notwithstanding, is 'damage' an essential element for establishing each type of GNSS civil liability? The answer is, beyond any doubt, yes. Unlike criminal law and regardless of tort liability or contractual liability, civil law does not impose civil liability for attempts. Even though numerous GNSS risks expose people's lives and property to danger (*see* §1.03), there is no civil liability for mere risk. In tort law, even though the extent of risk could affect the content of the civil liability regime, such as special rules for ultra-hazardous activity where strict liability applies, if the defendant's wrongful acts or omissions could not have caused the damage claimed and the situation will continue unchanged for a reasonable period, there is no ground for holding the defendant liable in torts.[104]

In contract law, although all civil liability arises from breach of contract, damage may not always exist. This is also true in the context of GNSS, as the following hypothetical case illustrates:

> Airport A and GNSS provider B concluded a contract that B is to provide 24/7 PNT service for aircrafts landing in that airport, and performance parameters are listed in the contract. Due to the negligence of B's employee, the PNT service malfunctioned for 30 seconds. During that period, flight C based on the GNSS landing system was to land at Airport A.

In this case:

(i) If an *accident* occurred to flight C because of that GNSS malfunctioning, the damage is obvious: loss of life and property, decrease of reputation, and other intangible loss;[105]
(ii) If an *incident* occurred to flight C for the same reason and caused negative effects to the flight, airport or airline, although the damage is still identifiable to some extent, only the intangible loss is left;[106]

103. The crash of a navigation satellite in orbit may also cause the damage to the outer space environment, but this is not the damage caused by GNSS discussed in this book.
104. Ariel Porat & Alex Stein, *Tort Liability under Uncertainty* (Oxford University Press, 2001), at 102.
105. The term accident is defined as an occurrence in flight which caused casualties and loss of property. *See* Annex 13 to the Convention on International Civil Aviation.
106. The term incident is defined as an occurrence associated with the operation of an aircraft which affects or could affect the safety of operation. In other words, there is no actual damage that happened in the incident but just the symptom of *accident*. *Ibid.*

(iii) If flight C happened to be cancelled or the pilots landed manually for other reasons, where would the damage be? Even though GNSS provider B did violate its contractual duty, should it be held liable in the case where no damage occurred?

Returning to the general theory of contract law, whereas damage arising from contractual breach is usually presented by 'expectation loss', if the party does not suffer any loss of expectation or it cannot be proven, that party could then make a claim for 'reliance loss'.[107] While the former refers to the benefits, expected by party A from the performance of party B's obligation, which were prevented by the breach of contract committed by party B, the latter refers to the expense incurred by party A to perform the contract which was however wasted because of non-performance of party B.[108] In the case of (iii) above, even if GNSS provider B performed its obligation accordingly, flight C would neither benefit from that performance, nor would it increase its cost as it did not intend to use that GNSS signal or service during those thirty seconds. Therefore, neither expectation loss nor reliance loss were incurred in the case of (iii), meaning that no damage occurred there and that, in turn, no one should be held liable for that breach of contract, as it should be. Even though the pilots unwillingly landed the flight manually because they realised that the GNSS signal was lost, only 'nominal damages' would be considered by the court, provided the aircraft landed safely and calmly as usual.[109]

Therefore, to establish GNSS civil liability at least one type of damage listed above has to be proven in court, regardless of whether the basis is tort law or contract law. Not all breaches of contract cause damage and are subject to contractual liability in the context of GNSS.

[E] Causal Link: A Challenge to Prove Behand 'Middle Tools'

Can radio signals really *hurt* people *directly*? The answer is no. Unlike a tangible object, a radio signal is tasteless and colourless, the existence of which we are even unable to sense unless we use special devices. Technically speaking, satellite navigation signals are 'soft', and this is decisive in determining that those signals could never *hurt*, *hit* or *collide* with people and their properties.[110] That is why the causal link in the case of GNSS is usually not apparent. However, there is civil liability for the damage *caused* by

107. *See Anglia Television Ltd. v. Reed*, [1971] 3 All E.R. 690.
108. *See* Joseph Chitty & H. G. Beale, *Chitty on Contracts* (Sweet & Maxwell, Limited, 2008), at 26.
109. *Nominal damages* are awarded as a sum of money that can be spoken of, but which sum has no existence of quantity, such as two cents, in cases, *inter alia*, where there are no damages in fact are or can be proven. William Benjamin Hale, *Handbook on the Law of Damages* (Рипол Классик, 1896), at 25.
110. The only possible way to make people feel threaten by a signal itself may be the danger of electromagnetic radiation to an individual's health. Fortunately, radio signals routinely are usually within safety standards too minimal or insignificant to be an issue. *See Frequently Asked Questions about the Safety of Radiofrequency (RF) and Microwave Emissions from Transmitters and Facilities Regulated by the FCC*, https://www.fcc.gov/engineering-technology/electromagnetic-compatibility-division/radio-frequency-safety/faq/rf-safety#Q6, last accessed 2 June 2016.

those signals in the legal sense. The term 'caused' does not require direct physical contact or connection. Like damage caused by a wrong information service, such as defective software and aeronautical chart,[111] as long as the radio signal is one of the reasons for damage, we can say that a causal link exists. Another analogy could be that of a car accident caused by the failure of the brake system to work due to a defective design; this failure of the brake system links damage to the manufacturers, even though the *brakes* never *touched* the victims. Therefore, to be qualified as a causer of damage, the causal link between GNSS signals and defective information, brake failure and other issues has to be proven via the pathway of 'middle tools', such as vehicles in transportation, for example in the case of economic loss of banks caused by wrong GNSS timing data, computers. What GNSS civil liability addresses is the consequence of GNSS signal, rather than the signal or the middle tool itself (*see also* §3.03[D]).

However, the real challenge in the case of GNSS liability is to prove that the damage in question was actually *caused* by GNSS. On the one hand, as a member of a high technology family, GNSS is too technically complicated to be understood by most laypersons; on the other hand, GNSS signals update second by second and it is impossible to store them unless conduct monitoring is agreed in advance and records are kept. This technical nature determines the difficulty of proving that the failure of GNSS was the contributing factor or one thereof to the damage.[112] Furthermore, in an accident it is usually various accumulated factors coming together at one time, rather than just one element, which cause the accident, thus making it very hard to peel the failure of GNSS away from the other factors. The above is particularly true for a claim based on tort law, which has a substantial causality requirement, contrary to contract law wherein causality is seldom an issue.[113] This circumstance urges us to find a feasible way to rebalance the inequality in court between GNSS providers and victims. Although *causal link* is an essential element to establish GNSS civil liability, it does not mean that a link has to be proven by the claimants. The reversal of the burden of proof in GNSS tort law, *inter alia*, could be considered as a solution in cases concerning GNSS civil liability.[114] Also, taking into account the difficulty in proving the *specific* causal link in the case of multiple providers or multiple actors in one set of GNSS providers, it would be a good point if we address joint liability in the context of GNSS damage.

In addition, it is worthwhile repeating here that no direct causal link exists under GNSS civil liability if the damage is caused by wrong map data or navigation route provided, designed or computed by certain value-added service providers (*see* §2.02).[115]

111. *See Saloomey v. Jeppesen & Co.*, 707 F.2d 671 (2nd Cir. 1983).
112. *See* EC, *supra* note 29.
113. Omri Ben-Shahar & Ariel Porat, *Fault in American Contract Law* (Cambridge University Press, 2010), at xii.
114. Another challenge is to decide to what extent a GNSS factor contributes to the damage. For example, if GNSS data is only reference material for pilot, who is actually flying the aircraft, can we say the failure of GNSS makes the plane finally crash and not the pilot? The answers to questions like this are found in §3.03[D].
115. For a recent case on damage caused by wrong map data rather than GNSS signals or services, *see* http://abcnews.go.com/International/woman-drives-car-canadian-bay-gps-wrong-directions/story?id=39115061, last accessed 1 June 2016.

§2.05 INTERNATIONAL CHARACTERISTICS OF GNSS CIVIL LIABILITY

Through decades of development, GNSS now has become a truly global resource[116] determined by its global deployment, coverage, and application (*see* §1.02[E]). The technical global nature of GNSS has been transferred to the international characteristics of *GNSS governance* and *GNSS law*. Added to this, criminal liability is in general a matter of national autonomy,[117] and administrative liability shares this similar feature because of its close connection with public powers originating from State sovereignty. Civil liability, however, can be engaged in both domestic law and governing rules of international law, especially for global systems such as GNSS and international air transportation:

(i) *Transnational litigant parties*. GNSS technology is relevant to all States regardless of their stages of economic or technological development.[118] This means that GNSS users and relevant third parties, who are potential claimants for compensation, are geographically distributed in every corner of the globe. At present, however, potential defendants, i.e., GNSS providers, can be counted on one's hand, therefore claimants do not have much choice but to aim at those few GNSS oligarchs for compensation. This would create a lot of transnational litigation with either foreign claimants or foreign defendants. Furthermore, current GNSS oligarchs enjoy and are more or less connected to State sovereignty (*see* §1.03, §1.04[C] and §5.03[B][2]), and usually this doctrine cannot be excluded merely by regulations under *national* law.

(ii) *Worldwide triggers of civil liability*. GNSS is a global business with its provision of a Global/World Public Good (GPG) (open signals),[119] so the triggers of civil liability are ready to be activated in one jurisdiction or in multiple jurisdictions. Even though the places where contracts have been signed would be limited to provider States within a certain number, the places where contracts have been performed and breached would be quite complicated in the context of GNSS, in particular for the regularly mobile vehicles; the malfunction of GNSS would cause damage to many targets all over the world which would lead to unlimited *locus delicti* and *lex loci*. The same applies to the product liability regime, taking into account the global supplier of GNSS satellites, software, and the signals transmitted to all corners of the world. The above unstable places of contract and tort makes it impossible to achieve the same result of choice of law even in the case of the same accident brought before the same court.

---

116. Someswar, et al., *supra* note 3.
117. Hui Wang, *Civil Liability for Marine Oil Pollution Damage: A Comparative and Economic Study of the International, US and the Chinese Compensation Regime* (Kluwer Law International, 2011), at 39.
118. Scott Madry, *Global Navigation Satellite Systems and Their Applications* (Springer, 2015), at 99.
119. *See* Serge Plattard, *Can Global Navigation Satellite Systems Signals Qualify to Become a World Public Good?*, 3(3) New Space 2015, at 142.

(iii) *Damage in multiple jurisdictions.* Modern GNSS has the technical ability to be used by an unlimited number of multimodal users at sea, on the ground and in the air,[120] with the result being that a single malfunction of GNSS would possibly damage each user at the same time. When the mobile feature of aircraft, vessels, and cars from one jurisdiction to another is taken into consideration, transboundary damage would be very common in case of accidents caused by GNSS.

Facing the challenges above, it is easy to imagine that the call for a global approach to GNSS civil liability has been the subject of extensive consultation in the international arena, including ICAO, UNIDROIT and IMO, for more than twenty years.[121] The lack of an international uniform instrument could cause many legal problems. First, transnational litigation would increase the difficulty and costs of identifying the liable party.[122] Second, the absence of a single criterion for private international law in the context of GNSS would expose the victims in the same event to the embarrassing situation of 'similar lives but with different values' due to the various indemnification standards of each State. Third, the cross-border damage means that jurisdiction-connecting factors related to GNSS accidents can range from the location of the damage to the location of the ground receiver stations, and this would inevitably cause conflict of jurisdictions and legal uncertainties.

Therefore, GNSS civil liability could to a significant extent be referred to as a type of international liability resulting from transboundary harm, in most situations because of its international characteristics. Like the legal regime of international air transportation,[123] a global approach has to be taken into consideration when discussing the establishment of a GNSS civil liability regime. It seems that only an international legal framework with mandatory effect could better ensure the equitable and uniform compensation for all affected persons, irrespective of the State to which they belong.[124] Being aware of the reluctance of States to reach a consensus on a new convention,[125] there is still a long way to go to find an acceptable global approach where alternative forms, such as unification of private international law rules, model law or guidelines, should also be on the list for consideration.

120. There are two different modes of GNSS technology, one is active systems, and the other is passive systems. Depending on the different modes, systems have a limited capacity or they serve an unlimited number of users. Currently, GPS, GLONASS, and Galileo are passive systems, and BDS keeps both active and passive methods. *See* Bernhard Hofmann-Wellenhof, et al., *GNSS – Global Navigation Satellite Systems* (Springer Science & Business Media, 2007), at 55 & 397.
121. Bollweg, *supra* note 38.
122. Manzini & Masutti, *supra* note 37.
123. GNSS civil liability seems to be more *international* than air carrier's civil liability: the international characteristic of air law is from the multinational victims' perspective but the damage usually happens in the same State; the international feature of GNSS civil liability is derived from both the multinational victims' perspective and the place of damages.
124. *See* Carbone & Maestri, *supra* note 81, at 41.
125. Francis P. Schubert, *An International Convention on GNSS Liability: When Does Desirable Become Necessary?*, XXIV Annals of Air and Space Law 1999, at 267.

§2.06 CONCLUDING REMARKS

It is not easy to define civil liability in the context of GNSS since few pre-existing reference materials focus specifically on this term. Following the general theory of liability, GNSS could trigger civil liability and criminal liability, as well as administrative liability, even though GNSS civil liability would be the major concern in most cases. Nevertheless, in the regime of international law, the demarcation between responsibility and liability in the context of GNSS should not be vague due to their duality in different areas of law.

Bearing those points in mind, GNSS civil liability could be defined as 'the obligation to make reparation for any damage caused, especially in the form of monetary payment, by the inappropriate PNT signal provided by core GNSSs, augmentation systems and regional systems, but excluding the damage caused by GNSS value-added services and the malfunction of user equipment'. The complete picture of GNSS civil liability is composed of contractual liability and non-contractual liability, and the largest percentage of the latter is represented by GNSS tort liability, even though it is not the entire picture. Both GNSS third-party liability and GNSS product liability are two subsets of GNSS tort liability which sometimes overlap.

To establish civil liability using the four elements structure, GNSS needs to identify the parties, the trigger, the damage, and the causal link. Considering the difficulty in determining the trigger and proving a causal link, specific legal arrangements such as the reversal of the burden of proof and joint liability must however be addressed more than they are in other sectors. More importantly, transnational litigant parties, cross-border triggers and damage in multiple jurisdictions determine the necessity of a truly global approach to deal with civil liability in the context of GNSS. Therefore, the following two chapters address whether current international aerospace (air and space) law could grapple with the overall situation of GNSS civil liability.

CHAPTER 3

International Space Law for GNSS Civil Liability: The Ideal Versus the Real*

§3.01 INTRODUCTION

Technology will never stand still,[1] even though no technological development ever occurs in a legal vacuum. GNSS is no exception.[2] In view of the fact that law is a maze rather than a motorway,[3] we are compelled to find appropriate law and to apply it in a positive manner. Old laws may fit new technology such as GNSS, but a legal gap between the ideal and the real cannot be denied since most pre-existing laws did not anticipate, when stakeholders were fighting intensely for their adoption, the essential role, and, in particular, the risks of GNSS (*see* §1.03). A careful review is necessary to determine whether current civil liability regimes, from an international perspective, can respond properly to the ongoing challenges in the GNSS era, especially challenges arising from the international character of GNSS civil liability (*see* §2.05).

As GNSS is a key element of space systems, international space law, among other branches of modern international law such as in terms of aviation (*see* Chapter 4), merits the first attempt at a response to the challenges of GNSS. Therefore, this chapter first aims to link GNSS with the legal sources under international space law (*see* §3.02), then to ascertain the actual relationship between those sources and GNSS civil liability by basically answering the following two key questions (*see* §3.03 and §3.04): (i) whether the liability regime for damage caused by space objects applies to GNSS

* Part of this chapter was published as: Dejian Kong, *International Space Law for GNSS Civil Liability: A Possible Solution?*, Space Policy, https://doi.org/10.1016/ j.spacepol.2019.01.001.

1. Bruce A. Hurwitz, *State Liability for Outer Space Activities in Accordance with the 1972 Convention on International Liability for Damage Caused by Space Objects* (Martinus Nijhoff Publishers, 1992), at 18.
2. Kim Murray, *The Law Relating to Satellite Navigation and Air Traffic Management Systems-A View from the South Pacific*, 53(2) Journal of Navigation 2000, at 385.
3. *Morris v. CW Martin and Sons Ltd* [1966] 1 QB 716, at 730.

damage; and (ii) if that regime applies, whether it is an adequate and appropriate mechanism for GNSS civil liability. This chapter concludes with some closing remarks (*see* §3.05).

§3.02 GNSS UNDER INTERNATIONAL SPACE LAW

[A] Overview of International Space Law

As the name implies, space law is the law that regulates space-related activities.[4] Space activities commonly occur in an international domain[5] and those parts of space law thus fall within international law,[6] thereby leading to the term 'international space law'. International space law may be nevertheless ambiguous when applying its general rules on responsibility and liability to such specific space activities as satellite remote sensing and satellite navigation.[7] It appears that only few legal documents and provisions of international space law address legal issues of satellite navigation.[8] Therefore, we first need to examine whether *the provisions of GNSS services or signals* could be qualified as 'space activities', and only then apply international space law.[9]

[B] The Term 'Space Activity' in the Context of GNSS

The term 'space activity' frequently appears in treaties, domestic legislation, and academic papers, although its specific definition is seldom found.[10] Yet, by referring to the definition of space law,[11] the wording of outer space treaties,[12] and relevant international documents,[13] the author believes that it is reasonable to define the term 'space activity' as follows:

---

4. UNOOSA, *Space Law*, http://www.unoosa.org/oosa/ourwork/spacelaw/, last accessed 2 May 2017.
5. Armel Kerrest & Caroline Thro, *Liability for Damage Caused by Space Activities*, in Ram S. Jakhu & Paul Stephen Dempsey (eds), Routledge Handbook of Space Law (Routledge, 2017), at 59.
6. Gabriella Catalano Sgrosso, *International Space Law* (LoGisma editore, 2011), at 27.
7. Assuyo Ito, *Legal Aspects of Satellite Remote Sensing* (Koninklijke Brill NV, 2011), at 244.
8. All the United Nations treaties, principles, and related General Assembly resolutions on Outer Space do not mention satellite navigation directly. *See* UNOOSA, *United Nations Treaties and Principles on Outer Space and Related General Assembly Resolutions* (United Nations, 2008), ST/SPACE/11/Rev.2.
9. Although the question when a GNSS *signal* is a space object remains further discussion below, it seems clear that GNSS *service* is not within the scope of space *object*.
10. For example, Section 103 of the US National Aeronautics and Space Act of 1958:

> the term 'aeronautical and space activities' means (A) research into, and the solution of, problems of flight within and outside the Earth's atmosphere, (B) the development, construction, testing, and operation for research purposes of aeronautical and space vehicles, (C) the operation of a space transportation system including the Space Shuttle, upper stages, space platforms, and related equipment, and (D) such other activities as may be required for the exploration of space.

11. For example, Francis Lyall & Paul B. Larsen, *Space Law: A Treatise* (Ashgate, 2009), at 2:

all human activates for the purpose of exploration of outer space including the Moon and other celestial bodies, and it ranges from the research, development, manufacture, operation and use of space infrastructure.

Furthermore, GNSS is a space-based system[14] and a space asset,[15] and its characteristics are similar to those of satellite remote sensing systems and satellite communication systems. Even though the operation activities are purely terrestrial undertakings, this does not reduce the space-based characteristics of a space system.[16] Therefore, GNSS-related activities ranging from launching navigation satellites to operating the whole navigation system qualify as 'space activities' as defined above, and thus render international space law applicable. This line of reasoning can be confirmed by the fact that most GNSS powers incorporate activities associated with satellite navigation to the authorisation and supervision scope of national space

---

> At its broadest space law comprises all the law that may govern or apply to outer space and activities in and relating to outer space.

12. For example, Article I of the Outer Space Treaty:

> Outer space [...] shall be free for exploration and use by all States [...] in accordance with international law.

Article III of the Outer Space Treaty:

> States Parties to the Treaty shall carry on activities in the exploration and use of outer space, including the moon and other celestial bodies, in accordance with international law.

13. For example, OECD, *Handbook on Measuring the Space Economy* (OECD Publishing, 2012), at 19:

> The space sector includes all actors involved in the systematic application of engineering and scientific disciplines to the exploration and utilisation of outer space, an area which extends beyond the earth's atmosphere.

14. GNSS was defined as follows in UNISPACE III Report:

> Global navigation satellite system (GNSS) are space-based radio positioning systems that provide 24-hour three-dimensional position, velocity and time information, in any weather conditions, to suitably equipped users anywhere on the surface of Earth, as well as airborne and space users.

UN, Report of the Third United Nations Conference on the Exploration and Peaceful Uses of Outer Space (Vienna, 19-30 July 1999), 18 October 1999, A/CONF.184/6, at 49.

15. Joan Johnson-Freese, *Space Warfare in the 21st Century* (Routledge, 2017), at 5.
16. For example, 'remote space activities' is defined as "the operation of remote sensing space systems, primary data collection and storage stations, and activities in processing, interpreting and disseminating the processed data" and although all the above activities are ground-based, this does not run counter to the nature of space-based systems of satellite remote sensing systems as they fall within the reach of 'Principles Relating to Remote Sensing of the Earth from Outer Space', which is an important element of international space law. *See* Principle I of the Principles Relating to Remote Sensing of the Earth from Outer Space, adopted by the United Nations General Assembly in its resolution 41/65 of 3 December 1986; Fabio Tronchetti, *Legal Aspects of Satellite Remote Sensing*, in Frans von der Dunk & Fabio Tronchetti (eds), Handbook of Space Law (Edward Elgar Publishing, 2015), at 520.

agencies, as well as by the fact that GNSS constitutes one of the competence items of UNOOSA.[17] Further, GNSS application was recognised as one of the great space applications by UNISPACE III.[18] However, 'merely receiving signals or information in some other form from objects in outer space is not designated as space activities',[19] and this is also consistent with the argument which excludes liability for malfunction of user equipment from the GNSS civil liability regime in this book (*See* §2.02).

[C] Sources of International Space Law Concerning GNSS Civil Liability

Compared with other branches of international law such as the law of the sea, international space law – the body of law governing space-related activities[20] – is much younger and has less legal instruments since the 'space age' is but a recent happening. The development of space law has been deadlocked for several decades since the Cold War. Currently, only the five space treaties, with legal binding effect, address the issue of fundamental rules on the exploration of outer space, namely the 1967 Outer Space Treaty,[21] the 1968 Rescue Agreement,[22] the 1972 Liability Convention,[23] the 1975 Registration Convention,[24] and the 1979 Moon Agreement.[25] Nevertheless, international liability was placed in quite an important position in the drafting history of the Outer Space Treaty and the Liability Convention,[26] and the positions of the Soviet Union and the US were unusually aligned on whether to include the issue of responsibility and liability in the discussion, even though their positions on many other issues were widely divergent.[27]

17. *See* UNOOSA, *Our Work*, http://www.unoosa.org/oosa/en/ourwork/index.html, last accessed 14 September 2017.
18. *The Space Millennium: Vienna Declaration on Space and Human Development*, adopted by the Third United Nations Conference on the Exploration and Peaceful Uses of Outer Space (UNISPACE III) at its 10th plenary meeting, 30 July 1999; Nie Jingjing, *The Future of Uniform International Rules on GNSS Liability*, 54 Proceedings of the International Institute of Space Law 2011, at 339.
19. Section 1 of the Swedish Act on Space Activities.
20. UNOOSA, *supra* note 4.
21. Treaty on Principles Governing the Activities of States in the Exploration and Use of Outer Space, including the Moon and Other Celestial (Outer Space Treaty), done 27 January 1967, entered into force 10 October 1967; United Nations, *Treaty Series*, Vol. 610, No. 8843.
22. Agreement on the Rescue of Astronauts, the Return of Astronauts and Return of Objects Launched into Outer Space (Rescue Agreement), done 22 April 1968, entered into force 3 December 1968; United Nations, *Treaty Series*, Vol. 672, No. 9574.
23. Convention on International Liability for Damage Caused by Space Objects (Liability Convention), done 29 March 1972, entered into force 1 September 1972; United Nations, *Treaty Series*, Vol. 961, No. 13810.
24. Convention on Registration of Objects Launched into Outer Space (Registration Convention), done 14 January 1975, entered into force 15 September 1976; United Nations, *Treaty Series*, Vol. 1023, No. 15020.
25. Agreement Governing the Activities of States on the Moon and Other Celestial Bodies (Moon Agreement), done 18 December 1979, entered into force 11 July 1984; United Nations, *Treaty Series*, Vol. 1363, No. 23002.
26. Stephan Hobe, Bernhard Schmidt-Tedd & Kai-Uwe Schrogl (Eds), *Cologne Commentary on Space Law: Volume I* (Carl Heymanns Verlag, 2009), at 130.
27. UNOOSA, *LSC Summary Records – 7th Session*, 1962, A/AC.105/C.2/SR.1.

As the 'constitution' of outer space, the Outer Space Treaty lays down the basic regulations and framework of outer space law, including liability for damage caused by space objects. Article VII of the Outer Space Treaty provides the legal basis to international claims for compensation,[28] and states that each *launching State* (*see* (v) of §3.04) shall be:

> internationally liable for damage to another State Party to the Treaty or to its natural or juridical persons by such object or its component parts on the Earth, in air space or in outer space, including the Moon and other celestial bodies.

The Liability Convention establishes specific provisions and categories of liability for space activities. It does so with reference to its legislative basis - Article VII of the Outer Space Treaty - which was considered to be insufficiently explicit about liability beyond its general provisions.[29]

In addition, the international space law community recognises the inherently ultra-hazardous nature of space activities.[30] As a result, strict/absolute liability is applied for damage on the surface of the Earth or in the aircraft in flight,[31] thereby excluding claimants from sharing the burden of proof of fault, to favour the interests of victims; fault-based liability regime is however applied for damage being caused in outer space.[32]

Liability under the Outer Space Treaty and the Liability Convention is geographically and financially unrestricted, and it provides maximum protection to potential victims.[33] Moreover, these two treaties impose international liability squarely and only on those States which qualify as launching States,[34] which ensures an efficient scheme for identifying the liable party and defendant. In addition, the Liability Convention elaborates relevant definitions, settlement of disputes procedure, joint liability regime, and other specific elements.

This victim-oriented civil liability regime established by both the Outer Space Treaty and the Liability Convention constitutes legal sources to analyse the matter of GNSS civil liability in international space law. Furthermore, since increasing State practice has seemingly furnished the necessary *opinio juris*, the author supports the viewpoint which asserts the status of customary international law of Article VII of the Outer Space Treaty,[35] in which case non-Member States could also apply a civil liability regime based on this provision.[36]

---

28. Hobe, Schmidt-Tedd & Schrogl, *supra* note 26, at 142.
29. *Ibid.*, at 136.
30. Isabella Henrietta Philepina Diederiks-Vershoor & Vladimír Kopal, *An Introduction to Space Law* (Kluwer Law International, 2008), at 37; *Ibid.*, at 143-144.
31. Article VII of the Outer Space Treaty; Article II of the Liability Convention.
32. Article III of the Liability Convention.
33. Hobe, Schmidt-Tedd & Schrogl, *supra* note 26, at 136.
34. Article VII of the Outer Space Treaty; Article II of the Liability Convention.
35. Ram S. Jakhu & Steven Freeland, *The Relationship Between the Outer Space Treaty and Customary International Law*, 59 Proceedings of the International Institute of Space Law 2016, at 183.
36. It should be noted here that (i) the status of customary international law only increases the scope of application to non-Member States, and does not affect the applicability of the Outer Space Treaty to damage caused by GNSS; (ii) the question of whether the status of customary

Scholars generally hold that the source of international space law should be the same as general international law.[37] Besides international treaties, the sources of international law recognised by the International Court of Justice (ICJ) also include: (i) customary international law, (ii) general principles of law, and (iii) works of highly qualified publicists.[38] Items (i) and (ii) are usually presented by general practice and national legal systems; Item (iii) works as subsidiary means for the determination of rules of law. Furthermore, soft law in outer space has also played an important role from the very beginning of space activities to the present.[39] Even though the non-binding characteristic of soft law in outer space does not always make it useful or appropriate for all international law jobs,[40] it at least can be used as a subsidiary means for settlement of disputes and supporting materials for the interpretation of conventions on international space law.

§3.03 IDEALISTIC: A POSSIBLE WAY TO APPLY INTERNATIONAL SPACE LAW TO GNSS CIVIL LIABILITY

[A] Arguments on the Applicability of International Space Law to GNSS Civil Liability

In reply to the question of whether the civil liability regime of international space law, i.e., Article VII of the Outer Space Treaty and the Liability Convention,[41] can apply to damage caused by GNSS, two opposing views exist in academia. One holds that the

international law of Article VII of the Outer Space Treaty extends to the Liability Convention remains open and needs further discussion, but is outside the scope of this book.

37. Article 38(1) of the Statute of the International Court of Justice, annexed to the UN Charter; Nicolas Mateesco Matte, *Space Activities and Emerging International Law* (McGill University, 1984), at 74; Qizhi He, *Outer Space Law* (Law Press China, 1992), in Chinese, at 21.
38. Article 38 of the Statute of the International Court of Justice, annexed to the UN Charter.
39. Irmgard Marboe (Ed.), *Soft Law in Outer Space* (Böhlau Verlag Wien Köln Graz, 2012), at 5.
40. William Michael Reisman, *Soft Law and Law Jobs*, 2(1) Journal of International Dispute Settlement 2011, at 25.
41. Before any further discussion, it should be noted that the author holds that what the Outer Space Treaty and the Liability Convention regulate are *civil* liability, as opposed to *administrative* liability arising from maladministration or negligence of supervision and regulation (*see* §2.03[B]). The reason is that: to favour the interest of victims, Article VII of the Outer Space Treaty and the Liability Convention require the State to be liable for the compensation resulting from the activities of its nationals, regardless of whether or not that State is liable for maladministration. State liability in international space law is in nature a vicarious liability for the civil damage caused by space objects based on private law theory, and therefore such national space legislation as the Swedish Act on Space Activities (Section 6) allow a State which has been designated for reimbursement from the persons who have carried on the space activity. Even the nature of State liability under the Liability Convention is civil liability; claimants still could not ask for compensation directly to the liable State based on the Liability Convention, as this Convention is for the claim presented by a State.

    However, an opposing opinion does exist, which distinguishes the terms 'State liability' and 'civil liability' and advocates that the liability regime under the Liability Convention is not civil liability. *See* OECD Nuclear Energy Agency, *Liability and Compensation for Nuclear Damage: An International Overview* (OECD Publishing, 1994), at 10.

current space law regime does offer civil remedy to GNSS damage.[42] The other rules out the applicability of the current space law regime to GNSS civil liability.[43] The author can neither support nor oppose either of these views since international space law can cover GNSS civil liability in a political sense, rather than legal.

Indeed, the core meaning of Article VII of the Outer Space Treaty could be construed as *launching States bear international liability for damage caused by space objects*, and the validity of this conclusion is supported by the full title of the Liability Convention - Convention on International Liability for Damages Caused by Space Objects - where the core words are 'damage caused by space objects'. Therefore, applicability depends on the understanding or interpretation of the phrase 'damage caused by space objects'. Consequently, the notions 'damage' (*see* §3.03[C]), 'caused by' (*see* §3.03[D]) and 'space object' (*see* §3.03[B]) are discussed individually so as to seek the possibility of applying international space law to GNSS civil liability.

[B] Definition of 'Space Object'

At the time when the Liability Convention was concluded there existed no generally accepted legal definition of the term 'space object'.[44] Although Article I of the Liability Convention lays down that "the term 'space object' includes component parts of a space object as well as its launch vehicle and parts thereof," most scholars recognise this provision as an *expression* rather than a full definition of space object,[45] or as a partial definition.[46] From a legal standpoint, it is without question that navigation satellites including their components are space objects here,[47] but it is questionable whether a *signal* transmitted by those navigation satellites could fall within the term 'space object' according to that expression.

For the above question, there are three schools of thought about this confusion. The first school insists that a space object itself would have both material and physical

42. Pablo Rodriguez-Contreras Perez, *GNSS Liability Issues: Possible Solutions to a Global System* (McGill University, 2002), at 58.
43. Ingrid Lagarrigue, *Are Existing Navigation Satellite Liability Provisions Adequate to Govern Navigation Satellite Malfunction*, 3(1) Outer Space Committee Newsletter 2000, at 31-32.
44. Elena Carpanelli & Brendan Cohen, *Interpreting 'Damage Caused by Space Objects' under the 1972 Liability Convention*, 56 Proceedings of International Institute of Space Law 2013, at 29.
45. Bin Cheng, *International Responsibility and Liability for Launch Activities*, 20(6) Air and Space Law 1995, at 297; Hobe, Schmidt-Tedd & Schrogl, *supra* note 26, at 139-140; Stephan Hobe, Bernhard Schmidt-Tedd & Kai-Uwe Schrogl (Eds), *Cologne Commentary on Space Law: Volume II* (Carl Heymanns Verlag, 2013), at 110 and 115; William F. Foster, *The Convention on International Liability for Damage Caused by Space Objects*, 10 The Canadian Yearbook of International Law 1972, at 144-145; Carl Q. Christol, *The Modern International Law of Outer Space* (Pergamon Press, 1982), at 108; Ra Michael Chatzipanagiotis & Konstantina Liperi, *Regulation of Global Navigation Satellite Systems*, in Ram S. Jakhu & Paul Stephen Dempsey (eds), Routledge Handbook of Space Law (Routledge, 2017), at 165; Boakye Danquah Kofi Henaku, *The International Liability of the GNSS Space Segment Provider*, XXI (1) Annals of Air and Space Law 1996, at 165.
46. Stephen Gorove, *Toward a Clarification of the Term 'Space Object': An International Legal and Policy Imperative?*, 21(1) Journal of Space Law 1993, at 12.
47. Cheng, *supra* note 45, at 297-310.

properties which excludes a signal.[48] The second school argues that damage from intangible electromagnetic waves was not absolutely excluded in the interpretation of the Liability Convention even though physical damage caused by tangible parts of a space object was of foremost concern.[49] The third school, while not popular with scholars, directly recognises that the signal emitted from the space object is indeed a space object.[50] Therefore, a proper interpretation to determine whether a space object must be tangible or material is essential in applying GNSS civil liability cases to the Outer Space Treaty and the Liability Convention.

Unlike the United Nations Convention on the Law of the Sea[51] and the Chicago Convention (*see* §4.04[C]),[52] the treaties on outer space themselves were not intended to be a permanent code but, in a more modest way, they were intended to articulate principles.[53] Further, they neither offer a mechanism for the interpretation of their provisions in general, nor do they design a remedy for the settlement of disputes in understanding those provisions.[54] In this case, we have to make reference to the rules of interpretation laid down in general international law: Articles 31-33 of the Vienna Convention on the Law of Treaties (Vienna Convention)[55] which represents the final and authoritative achievement of decades of efforts on treaty interpretation.[56]

The interpretation rules set down by the Vienna Convention stipulate that a treaty must be interpreted by the *ordinary meaning* of its terms with reference to *the context*

---

48. Carl Q. Christol, *International Liability for Damage Caused by Space Objects*, 74 American Journal of International Law 1980, at 354; Michael Milde, *Solutions in Search of a Problem? Legal Problems of the GNSS*, XXII(2) Annuals of Air and Space Law 1997, at 212; Hobe, Schmidt-Tedd & Schrogl, *supra* note 26, at 139; Kerrest & Thro, *supra* note 5, at 64.
49. Lesley Jane Smith, *Legal Aspects of Satellite Navigation*, in Frans von der Dunk & Fabio Tronchetti (eds), Handbook of Space Law (Edward Elgar Publishing, 2015), at 585.
50. Henaku, *supra* note 45, at 165.
51. Part XV of the United Nations Convention on the Law of the Sea.
52. Article 84, Chapter XVIII of the Chicago Convention.
53. Secure World Foundation, *Outer Space Treaty Fiftieth Anniversary*, https://swfound.org/media/205736/ost50_transcript_jan_2017.pdf, last accessed 5 June 2017.
54. Hanneke van Traa-Engelman, *Settlement of Space Law Disputes*, 3(3) Leiden Journal of International Law 1990, at 139-155.
55. Here may arise a question of how the interpretation of a preceding treaty, for example, the Outer Space Treaty taking effect in 1967, apply rules codified by a later treaty, the Vienna Convention taking effect in 1980. For this question, certain scholars have already made a convincing point of view and case analysis by the following words:

    > The International Court of Justice (ICJ) has on several occasions confirmed that both Article 31 and Article 32 of the Vienna Convention reflect customary international law and has applied these rules to treaties that predated the Vienna Convention. For example, in 1999, the Court interpreted and applied the rules codified in Article 31 and 32 of the Vienna Convention, when considering the meaning of a treaty was concluded in 1890.

    Ram S. Jakhu & Steven Freeland, *The Relationship Between the United Nations Space Treaties and the Vienna Convention on the Law of Treaties*, 55 Proceedings of the International Institute of Space Law 2012, at 386-387.
56. Ulf Linderfalk, *Is the Hierarchical Structure of Article 31 and 32 of the Vienna Convention Real or Not? Interpreting the Rules of Interpretation*, 54 (1) Netherlands International Law Review 2007, at 134; Evan Criddle, *The Vienna Convention on the Law of Treaties in U.S. Treaty Interpretation*, 44(2) Virginia Journal of International Law 2004, at 433.

and *the object and purpose* thereof.[57] In other words, the following three primary means of interpretation that can be used by an interpreter citing Article 31 are (i) conventional language, (ii) the context, and (iii) the object and purpose of a treaty:[58]

(i) *Conventional language.* Determining the ordinary meaning of the term 'object' is the point of departure for understanding the term 'space object' used by the Outer Space Treaty and the Liability Convention. This term is in daily usage,[59] and it usually refers to a material thing that can be seen and touched with a fixed shape or form.[60] However, non-tangible radiations, where GNSS signals are included, are a series of radio waves with electronic information,[61] and hence they are not even 'objects', let alone 'space objects'.[62] In addition, although the author agrees that Article 1(d) of the Liability Convention does not qualify as a definition of the term 'space object', it indeed may serve as a basis to understanding the meaning of that term.[63] The non-definition is the result of the Legal Sub-committee of UNOOSA believing that the term 'space object' had a reasonably clear meaning and it was only necessary to emphasise that all the component parts and launching devices were included besides a space object itself.[64] In this sense, the minimum requirement of a space object is a physical nature, otherwise no component parts or launching devices thereof could be included.

(ii) *The context.* The terms of a treaty are not drafted in isolation, and we must consider their normal meaning within the entire treaty text.[65] In the context of Article VII of the Outer Space Treaty and the Liability Convention, phrases such as 'the launching of an object to outer space',[66] 'a space object is launched',[67] 'launch a space object',[68] and 'the operation of that space

57. Article 31 of the Vienna Convention reads as follows:

> A treaty shall be interpreted in good faith in accordance with the ordinary meaning to be given to the terms of the treaty in their context and in the light of its object and purpose.

58. Linderfalk, *supra* note 56, at 153.
59. Gorove, *supra* note 46, at 25.
60. *Collins COBUILD Advanced Learner's Dictionary* (Harper Collins Publishers, 2001), at 1058; *The Concise Oxford Dictionary* (Foreign Language Teaching and Research Press & Oxford University Press, 1999), at 938; *Longman Dictionary of Contemporary English* (Foreign Language Teaching and Research Press, 2002), at 973; *Webster's New World College Dictionary* (Liaoning Education Press & Hungry Minds Inc.), at 994; *Macmillan English Dictionary for Advanced Learners of American English* (Foreign Language Teaching and Research Press, 2002), at 958.
61. Collins, *supra* note 60, at 1448; Oxford, *supra* note 60, at 1291; Longman, *supra* note 60, at 1335; Webster's, *supra* note 60, at 1333; Macmillan, *supra* note 60, at 1306.
62. Gorove, *supra* note 46, at 25.
63. Hobe, Schmidt-Tedd & Schrogl, *supra* note 45, at 115.
64. Foster, *supra* note 45, at 145; UNOOSA, *LSC Summary Records – 7th Session*, 1968, A/AC.105/C.2/SR.106.
65. Mark E. Villiger, *Commentary on the 1969 Vienna Convention on the Law of Treaties* (Martinus Nijhoff Publishers, 2009), at 427.
66. For example, Article VII of the Outer Space Convention.
67. For example, Article I of the Liability Convention.

object' are frequently used, and this seems that a space object is usually connected with 'launching' and 'operating' activities (*see* (v) of §3.04) even though those activities may not be essential for each space object.[69] This argument could also be supported by the academic definition of the term 'space object', which reads that "anything that human beings 'launch' or attempt to 'launch' into space".[70] More broadly, though still within the system of UN Treaties on Outer Space, the Registration Convention shares the same expression and meaning of space object word for word.[71] Article IV (d) thereof states that the basic orbital parameters including nodal period, inclination, apogee, and perigee concerning each space object shall be carried on the registry of each State, and in this sense it seems clear that the term 'space object' excludes GNSS signal as it has nothing to do with those orbital parameters. Another similar situation in the Moon Agreement is with the phrase such as 'land their space objects on the moon and launch them from the Moon'.[72] Of course, based on the analysis above, the author does not argue that each space object must be able to be 'launched', 'operated', 'registered', 'returned', and 'landed', but at least these expressions show a strong implication for the physical needs of a space object within the context of UN outer space treaties, particularly the Outer Space Treaty and the Liability Convention.

(iii) *The object and purpose.* The author agrees that the purpose of the civil liability regime under international space law, in particular the Liability Convention, is to ensure the prompt, adequate, and equitable compensation to victims for damage caused by space objects.[73] Based on this victim-oriented character, someone may argue or support the opinion that a broad interpretation to encompasses damage from 'intangible electromagnetic waves' would be reasonable.[74] The author does not share this view. The purpose of interpretation is to determine the original meaning of terms or provisions so that interpreters may not make new rules or revise the convention without the approval of all contracting States. It should be noted

68. For example, Article V of the Liability Convention.
69. Gorove, *supra* note 46, at 17-18.
70. Cheng, *supra* note 45, at 297.

Professor Vladimir Kopal also made a similar but a bit complicated definition to the term 'space object' as follows:

> As 'space object' should be considered any object launched by man for a mission into outer space, be it into orbit around the Earth or beyond/i.e. into interplanetary space, to and around the Moon and other celestial bodies of the Solar system, or into deep space.

Vladimir Kopal, *Some Remarks on Issues Relating to Legal Definitions of 'Space Object', 'Space Debris' and 'Astronaut'*, 37 Proceedings of the International Institute of Space Law 1999, at 101.

71. Article 1(b) of the Registration Convention.
72. For example, Article 8 of the Moon Agreement.
73. Preface of the Liability Convention.
74. Chatzipanagiotis & Liperi, *supra* note 45, at 165; Smith, *supra* note 49, at 585.

> that only when a particular treaty provision is ambiguous that an interpretation would be necessary. A treaty must be interpreted under the principle of good faith, and it would be inappropriate to 'read into' that provision certain rules so as to reflect what *should* be, particularly as such rules go beyond the normal meaning within the treaty context as required by Article 31.1 of the Vienna Convention.[75]

Literally speaking, it seems clear that a space *object* must be physical and hence excludes GNSS signal in its definition. However, if we look at Article 31.4 of the Vienna Convention which allows a *special* meaning of a treaty term, a question may arise as to whether it is possible to understand, in this way, that a non-material object, including a GNSS signal, was intentionally put into a special meaning of the term 'space object' by the drafters. The right answer to that question depends on whether "the parties so intended".[76] The intention to give an unusual meaning to a treaty term must be supported by direct evidence, in particular the *travaux préparatoires* which are the official records of a negotiation.[77] The past tense of the term 'intended', used in Article 31.4 of the Vienna Convention, also directs us to examine the historical materials as well.

Even though the Outer Space Treaty (Article VII), including its predecessor entitled 'Declaration of Legal Principles Governing the Activities of States in the Exploration and Use of Outer Space (Item 9)', and the Liability Convention finally selected the term 'object' or 'space object', the starting point in their draft documents submitted by individual Member States were such terms as (i) damage caused by 'space vehicles',[78] 'space devices', and 'the launching of objects into outer space',[79] and (ii) liability for a 'space vehicle accident'.[80] Those terms imply that what the delegations looked into was the civil liability for damage resulting from a physical

---

75. Jakhu & Freeland, *supra* note 55, at 387.
76. Article 31.4 of the Vienna Convention.
77. Richard K. Gardiner, *Treaty Interpretation* (Oxford University Press, 2015), at 70.
78. *See* UN Doc. A/4141, *Report of the Ad Hoc Committee on the Peaceful Use of Outer Space*, 14 July 1959, GA Official Records, Fourteenth Session, Agenda Item 25, Annexes, at 23; UN Doc. A/AC.105/C.2/L.4, *USA: Proposal-Liability for Space Vehicle Accidents*, 4 June 1962, in UN Doc. A/AC.105/6, Report of Legal Sub-Committee on the Work of its First Session (28 May-20 June 1962), 9 July 1962, at 6.
79. UN Doc. A/AC.105/C.2/L.8, *United States: Proposal-Convention Concerning Liability for the Launching of Objects into Outer Space*, 9 March 1964, in UN Doc. A/AC.105/19, Report of the Legal Sub-Committee on the Work of its Third Session (9-26 March 1964), 26 March 1964, Annex II, Proposals and amendments relating to liability for damage caused by objects launched into outer space, at 2; UN Doc. A/AC. 105/C.2/L.10, *Hungary: Proposed Draft Agreement – Agreement Concerning Liability for Damage Caused by the Launching of Objects into Outer Space*, 16 March 1964, in UN Doc. A/AC.105/19, Report of the Legal Sub-Committee on the Work of its Third Session (9-26 March 1964), 26 March 1964, Annex II, Proposals and amendments relating to liability for damage caused by objects launched into outer space, at 7.
80. *See* UNGA Res. 1802 (XVII), International cooperation in the peaceful uses of outer space, 14 December 1962, Article I para. 3; UN Doc. A/AC.105/35, Report of the Legal Sub-Committee on the Work of its Fifth Session (12 July – 4 August and 12 – 16 September 1966), 16 September 1966, at 2.

object itself, mainly in a space vehicle's accident,[81] particularly at the moment of launching,[82] rather than the intangible data, application or product emanating from that object.[83] The author found no evidence, in the historical context of international space law, which showed the drafters' intention to establish a connection between an intangible signal with a liability mechanism. Taking a step back, even though early GNSS – TRANSIT (*see* §1.02[A]) – had been in operation at the time of drafting the liability provisions of international space law, at the beginning of the space era large-scale application, particularly in such a safety-of-life field as aviation, was more like science fiction. The author therefore believes that there were few possibilities for the drafters, in particular of the Liability Convention, to even recognise the necessity to make civil liability regulations for an intangible GNSS signal.[84] The author would of course not deny the possibility of applying old law to new technology or situations,[85] but the above arguments show that no historical context supports the intention to add a special meaning to the treaty term 'space object'.

When taking a further step toward State practice, one scholar found that, while the majority of States do not define the term 'space object' in their national law, certain space powers simply copied the expression of space object from Article I of the Liability Convention, and only a few States give it a specific definition.[86] Similar to international treaties on outer space activities, national legislation, and academic viewpoints thereof also make the term 'space object' a collective term that includes 'space vehicle', 'spacecraft', 'spaceship', 'satellite', and 'space station' (*see (iii) The object and purpose*).[87] This scholar also concluded seven common elements of the definition of the term 'space object' in national laws, namely (i) object, (ii) intent to launch, (iii) launched, (iv) launch vehicle, (v) payload, (vi) physical component parts and parts thereof, and (vii) satellite.[88] In this scenario, the author does not see any major difference on the content of regulations between international treaties and State practice, regardless of whether or not those practices could be recognised as customary international law or general principles of law.

---

81. Christol, *supra* note 48, at 355; Roderick D van Dam, *GNSS and Aviation: Eurocontrol's Perspective*, Outer Space Committee Newsletter, 2000, at 48; Henaku, *supra* note 45, at 164.
82. Hobe, Schmidt-Tedd & Schrogl, *supra* note 45, at 102.
83. Stephen Gorove, *Some Thoughts on Liability for the Use of Data Acquired by Earth Resources Satellites*, 15 Proceedings of the International Institute of Space Law 1972, at 109; Hobe, Schmidt-Tedd & Schrogl, *supra* note 45, at 111.
84. Ruwantissa Abeyratne, *Space Security Law* (Springer, 2011), at 25; Lagarrigue, *supra* note 43, at 32.
85. The author agrees that old law could be applied to new technology but the key point is to see whether the new technology is merely a change in degree, an improved version of something that already exists, or a change in kind, something else entirely with a new capability. *See* Rebecca J. Rosen, *The Thorny Combination of Old Laws and New Tech*, https://www.theatlantic.com/technology/archive/2011/11/the-thorny-combination-of-old-laws-and-new-tech/248111/, last accessed 2 May 2017.
86. *See* Christopher M. Hearsey, *Comparative Study of the Definition of Space Object in National Space Laws and Its Legal Effect under International Law*, https://papers.ssrn.com/sol3/papers.cfm?abstract_id=2072514, last accessed 20 July 2017.
87. *Ibid.*
88. *Ibid.*

Therefore, here we can draw a simple conclusion that both international treaties on outer space, which are the Outer Space Treaty and the Liability Convention, and State practice, some of which may be recognised as customary international law or general principles of law, show no support for the viewpoint of interpreting and adding an intangible GNSS signal to the physical term 'space object'.

[C] Broad Interpretation of the Term 'Damage'

In order to apply the Liability Convention in the case of an accident caused by the failure of GNSS, within the context of 'damage caused by space objects', certain scholars try to interpret the term 'damage' broadly, which latter term is considered by academics as one of the most controversial aspects of legal history.[89] They incorporate the notion of *indirect damage*,[90] and hence argue that damage caused by GNSS could be recognised as indirect damage, which is covered by the Liability Convention.[91]

Even though the term 'damage' is clearly defined in Article I of the Liability Convention as "loss of life, personal injury or other impairment of health; or loss of or damage to property of States or of persons, natural or juridical, or property of international intergovernmental organizations," many scholars still insist that this definition is ambiguous, particularly in terms of whether that term includes only direct damage or, *inter alia*, indirect damage as well.[92] Indeed, during the drafting of the Liability Convention, the inclusion of indirect damage was such a thorny question that it did not result in an agreement.[93]

On the one hand, the US delegation expressed that the Liability Convention "does not cover what some delegations earlier called remote or indirect damage and for which there is only a hypothetical causal connection with a particular space activity,"[94] and pointed out that the question of indirect damage could cause great difficulties in

89. Anna Masutti, *GNSS: The Basic Principles for a European Legal Framework on TPL*, in Alfredo Roma, Kai-Uwe Schrogl & Matxalen Sanchez Aranzamend (eds), Policy Aspects of Third-Party Liability in Satellite Navigation (ESPI, 2009), at 33.
90. The typical case on 'indirect damage' in international space law is the crash of the USSR's Cosmos 954 Satellite, where Canada claimed the recovery of cleaning costs due to the nuclear contamination of vast stretches of Canadian territory. *See* Bryan Schwartz & Mark L. Berlin, *After the Fall: An Analysis of Canadian Legal Claims for Damage Caused by Cosmos 954*, 27 McGill Law Journal 1982, at 716.
91. Carpanelli & Cohen, *supra* note 44, at 45; Henaku, *supra* note 45, at 170; Chatzipanagiotis & Liperi, *supra* note 45, at 165.
92. Piotr Manikowski, *Examples of Space Damages in the Light of International Space Law*, 6(1) The Poznań University of Economics Review 2006, at 60; Andrzej Górbiel, *Outer Space in International Law* (Uniwersytet Łódzki, 1981), at 107; Carpanelli & Cohen, *supra* note 44, at 31; Andreas Loukakis, *Non-Contractual Liabilities from Civilian Versions of GNSS: Current Trends, Legal Challenges and Potential* (Nomos, 2017), at 29.
93. Report, A/AC.105/37, para. 17; Nandasiri Jasentuliyana & Roy S. K. Lee (Eds), *Manual on Space Law: Volume I* (Oceana Publications, 1979), at 115; UN, *Yearbook of the United Nations 1967* (United Nations, 1969), at 31; Kerrest & Thro, *supra* note 5, at 67; Carpanelli & Cohen, *supra* note 44, at 44.
94. Committee on Aeronautical and Space Sciences United States Senate, *Convention on International Liability for Damage Caused by Space Objects: Analysis and Background Data-Staff Report* (U.S. Government Printing Office Washington, 1972), at 24.

practice.[95] Certain scholars have supported this argument.[96] In the context of GNSS damage, the opinion based on the above position is also popularly accepted,[97] for example:

> Neither the language of the Convention, the negotiations leading to this Convention, nor State practice support such a claim [that the Liability Convention applies to indirect damage arising from the use of navigational satellite services].[98]

On the other hand, a few delegations, for example India,[99] were not satisfied with such a narrow interpretation as above, and certain scholars further support them as well.[100] They believe that the notion of damage in Article I of the Liability Convention generally covers both direct and indirect damage,[101] and only in this way could the Liability Convention live up to its victim-oriented nature. More importantly, many scholars share the above as specific to GNSS damage,[102] for example:

---

95. Committee on Aeronautical and Space Sciences United States Senate, Soviet Space Programs, *1966-70: Goals and Purposes, Organization, Resources, Facilities and Hardware, Manned and Unmanned Flight Programs, Bioastronautics, Civil and Military Applications, Projections of Future Plans, Attitudes Toward International Cooperation and Space Law. Staff Report* (U.S. Government Printing Office, 1971), at 481.
96. Kerrest & Thro, *supra* note 5, at 57; Valerie Kayser, *Launching Space Objects: Issues of Liability and Future Prospects* (Springer Science & Business Media, 2006), at 49; Marco Ferrazzani, *The Role and Liabilities of Space Segment Operators*, in European Centre of Space Law, Regulation of the Global Navigation Satellite System (GNSS): A Conference to examine Legal and Policy interests involved in the implementation of GNSS (ESTEC, 14-15 November 1996), at 160; Diederiks-Vershoor & Kopal, *supra* note 30, at 39; Frans von der Dunk, *International Space Law*, in Frans von der Dunk with Fabio Tronchetti, Handbook of Space Law (Edward Elgar Publishing Limited, 2015), at 84; Frans von der Dunk, *European Space Law*, in Frans von der Dunk with Fabio Tronchetti, Handbook of Space Law (Edward Elgar Publishing Limited, 2015), at 265; Paul B. Larsen, Joseph Sweeney & John Gillick, *Aviation Law: Cases, Laws and Related Sources* (Martinus Nijhoff Publishers, 2012), at 1052; Edward R. Finch, *Outer Space Liability: Past, Present and Future*, 14(1) The International Lawyer 1980, at 126; Gorove, *supra* note 83, at 109.
97. *See* Abeyratne, *supra* note 84, at 25; Lagarrigue, *supra* note 43, at 32; Francis P. Schubert, *An International Convention on GNSS Liability: When Does Desirable Become Necessary?*, XXIV Annals of Air and Space Law 1999, at 252; UNIDROIT, *An Instrument on Third Party Liability for Global Navigation Satellite System (GNSS) Services: A Preliminary Study* (UNIDROIT, 2010), Study LXXIX - Preliminary Study, March 2010, at 21; Larsen, Sweeney & Gillick, *supra* note 96.
98. Lagarrigue, *supra* note 43, at 32.
99. *See* UN Doc. A/AC.105/C.2/L.26, *India: Draft Agreement on Liability - Proposal*, 30 June 1967, in UN Doc. A/AC.105/37, Report of the Legal Sub-Committee on the Work of its Sixth Session (19 June - 14 July 1967), 14 July 1967, Annex II, Proposals, amendments and other documents relating to liability for damage caused by the launching of objects into outer space, at 20.
100. Lyall & Larsen, *supra* note 11, at 405; Hurwitz, *supra* note 1, at 15; Christol, *supra* note 48, at 362; Carpanelli & Cohen, *supra* note 44, at 39.
101. Hobe, Schmidt-Tedd & Schrogl, *supra* note 45, at 112; Carpanelli & Cohen, *supra* note 44, at 35.
102. Pablo Rodriguez-Contreras Perez, *Damage Caused by GNSS Signals in the Light of the Liability Convention of 1972*, in Michael Rycroft (eds), Satellite Navigation Systems: Policy, Commercial and Technical Interaction (Springer-Science + Business Media, B.V., 2003), at 252; Henaku, *supra* note 45, at 170; Chatzipanagiotis & Liperi, *supra* note 45, at 165.

> The conclusion that GNSS satellite damage other than collision is covered by the Liability Convention is not only correct from the reading of the provision but is also supported by the travaux préparatoires.[103]

To comment or make a choice between these two opposing arguments, the first thing needed is to understand what constitutes indirect damage in the context of space law. Indeed, the term 'indirect damage' is opposed to 'direct damage', but the distinction between them has been long criticised for its complexity and confusion, and case law states that there should be no place for the theory of indirect damage in international law.[104] Nevertheless, since the possibility to recognise an intangible GNSS signal as a space object was disconfirmed (*see* §3.03[B]), the author would like to discover whether the notion of indirect damage could be an alternative solution which is established on a different legal basis,[105] with the help of a hypothetical case model as follows:

> An aircraft with 300 passengers crashed into a farmer's house because the GNSS Landing Systems (autonomous Landing) broke down due to defective GNSS signals, and all the crew, passengers and the farmer lost their lives.[106]
>
> The defective GNSS signal resulted from the malfunction of GNSS satellites because of (i) their collision with a space object (A); (ii) the radio interference from a space object (A); or (iii) an accounting error, defective components and other defects of GNSS itself.

This case model could also be illustrated by Figure 3.1.[107]

---

103. Henaku, *supra* note 45, at 170.
It should be noted here that the citation here is not in conflict with the above argument that GNSS signals could not be recognised as space objects. What the *travaux préparatoires* supports here is that damage caused by a GNSS satellite (vs GNSS signal) is covered by the Liability Convention, and the key term here is 'damage' rather than 'space object'.
104. *See* Francisco V. Garcia Amador, Louis Bruno Sohn & Richard R. Baxter, *Recent Codification of the Law of State Responsibility for Injuries to Aliens* (Martinus Nijhoff Publishers, 1974), at 124; UN, *Report by Special Rapporteur of the International Law Commission (Arangio-Ruiz)*, UN Doc.A/CN.4/425, para. 36; UN, *Reports of International Arbitral Awards* (UN, 1956), at 62-63; Elihu Lauterpacht, Christopher J. Greenwood & Andrew G. Oppenheimer, *International Law Reports (117)* (Cambridge University Press, 2000), at 248.
105. *See* Hobe, Schmidt-Tedd & Schrogl, *supra* note 45, at 129.
106. The author finds a similar case in the context of satellite communication which is the response to the US delegation who explained that indirect damage does not apply the Liability Convention, and this case and opinion is quite helpful to the research. The original words are as follows:

> Only when damage results from this interference is the Convention applicable: thus, if for example a space object of one state interrupts the transmission of radio signals from a communications satellite to an aircraft in flight, which makes that aircraft veer off course and crash, the first-mentioned state may be held liable by virtue of article II of the Convention.

Peter van Fenema, *The 1972 Outer Space Liability Convention* (McGill University, 1973), at 62. The opinion which supports to apply the Liability Convention for damage caused by radio interference please *see also* Hurwitz, *supra* note 1, at 20.
107. In the pictures used in this chapter, all full lines refer to the fact that physical connect exists between heading and ending points, and by contrast all dotted lines mean that no physical connect exists.

Figure 3.1 Case Model

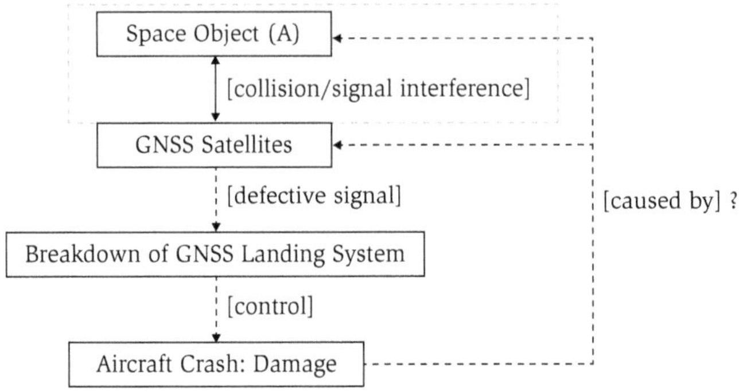

In this case model, it is not important what makes GNSS signals defective, but what is important is the fact that neither GNSS satellites nor Space Object (A) caused the damage with a direct and physical connection, which is not the usual way of 'damage caused by space objects'. Further, it is quite clear that GNSS satellites qualify as space objects regulated by Article VII of the Outer Space Treaty and the Liability Convention, and damage includes the personal casualty (passengers, crew, and the famer) as well as the loss of property (aircraft and house) in compliance with Article I (a) of the Liability Convention. Therefore, the key point here is not the question of whether indirect damage constitutes 'damage' as required by the Liability Convention, but whether we could say that damage is 'caused by' those GNSS satellites or Space Object (A), and hence the Liability Convention applies.

Regardless of the difficulty of finding an exact definition of the term 'indirect damage',[108] it is more or less right to say that it is the damage which is caused indirectly.[109] In this sense, the author holds that the nature of the question of whether 'damage indirectly caused by GNSS satellites' can be regarded as 'damage caused by space objects' depends more on the debate about the *causal link* between effect and activity and what degree of causality is required to bring about liability,[110] rather than

108. The notion of direct damage is emphasised from different perspectives in space law as well as the law of GNSS among international scholars. For example, Professor Smith and Professor Kerrest direct this notion to be "caused after an interval, an intervening event or events that are a consequence of the initial 'impact'"; Professor Masutti addresses indirect damage from the perspective of 'loss of profit'; Professor Mendes de Leon and Professor van Traa indicate that indirect damage, in the context of GNSS, refers to "damage caused by the signals in contrast with damage caused by the space object"; Dr Andreas Loukakis holds that indirect cases of damage are resulted from "the use of capabilities of a space object" rather than the space object as such, where damage caused by defective signals emitted by GNSS satellites qualify as a typical example. *See* respectively: Hobe, Schmidt-Tedd & Schrogl, *supra* note 45, at 127; Masutti, *supra* note 89, at 33; Pablo Mendes de Leon & Hanneke van Traa, *Space Law*, in Jessica Schechinger (eds), The Practice of Shared Responsibility in International Law (Cambridge University Press, 2017), at 475 (note 75); Loukakis, *supra* note 92, at 31.

109. Christol, *supra* note 48, at 360.

110. Nicolas Mateesco Matte, *Aerospace Law: From Scientific Exploration to Commercial Utilization* (The Carswell Company Limited & Editions A. Pedone, 1977), at 157.

on the pros and cons of the definition of damage (direct damage versus indirect damage) in outer space treaties. And this argument is at least supported by some delegates if we look into the *travaux préparatoires* of the Liability Convention.[111] For example, after repeating the uncertainty of the term 'indirect damage' in general international law and the case law of international arbitration, the Japanese delegation believed that:

> all damages which have an adequate relationship of cause and effect with the space activities should be covered in this convention. In order to avoid endless discuss on whether to include those terms of 'indirect damage' or 'delayed damage' in the definition of damage, we should discuss the problem of these two terms not in which the damage occurred, by introducing the notion of adequate relationship of cause and effect or so called 'the existence of proximity' in the Anglo-American laws.[112]

Although no text was finally added to clearly state that the Liability Convention covers indirect damage, it seems unclear whether this means that the delegations finally decided to exclude the applicability of damage caused indirectly, or whether they simply refused to accept Japan's proposal. Therefore, it is reasonable to argue that the difference between direct and indirect damage is a matter of adequate causation, which was not expressed in the Liability Convention.[113] A few scholars have extended this argument to the context of GNSS civil liability,[114] while the majority of scholars were still focusing on the definition of damage itself, i.e., whether GNSS damage could be included in the term 'damage' under the Liability Convention. The author shares the minority view and believes that, if the claim for GNSS damages intends to qualify under the outer space treaties, then the only matter to be proven lies in the *causation* between *damage* and *GNSS satellite/Space Object (A)* (*see* §3.03[D]),[115] particularly in the sense of the phrase 'damage caused by space objects'.

[D] The Matter of Causation

For the matter of causation, it is quite difficult to agree on a common definition in one specific convention where the conflict between *common law* and *civil law* has to be coordinated. That difficulty in turn leaves broad discretion for the dispute settlement body to identify that causation on a case-by-case basis in light of the intent and purposes of the convention, as well as by observing justice and equity.[116] Unlike the air law system where the exact meaning or test of causation is usually for domestic

111. *See* UNOOSA, *LSC Summary Records – 7th Session*, 1968, A/AC.105/C.2/SR.103.
112. UNOOSA, *Japan: Working Paper*, A/AC/105/C.2/L.61, 1969; Nandasiri Jasentuliyana & Roy S. K. Lee (eds), *Manual on Space Law: Volume III* (Oceana Publications, 1981), at 354.
113. Bin Cheng, *Studies in International Space Law* (Clarendon Press Oxford, 1997), at 323; Perez, *supra* note 42, at 46; Aldo Armando Cocca, *From Full Compensation to Total Responsibility*, 26 Proceedings of the Twenty-sixth Colloquium on the Law of Outer Space 1983, at 158.
114. For example, Perez, *supra* note 42, at 46 and 61; Perez, *supra* note 102, at 252.
115. *See* Hobe, Schmidt-Tedd & Schrogl, *supra* note 45, at 191.
116. Kayser, *supra* note 96, at 48-49.

tribunals to decide,[117] it seems that outer space law has to deal with the matter of causation in a more international sense, and the reason for this refers to two aspects. On the one hand, the outer space treaties do not give a general answer for the causation, but Article XII of the Liability Convention provides that the compensation "shall be determined in accordance with 'international law' and the principles of justice and equity."[118] On the other hand, claims under the Liability Convention must be based on the model of State-versus-State by a Claims Commission, rather than by a municipal court (*see* §3.04).[119]

Focusing on the context of GNSS civil liability, the matter of causation depends on the understanding of the term 'caused by' under the phrase 'damage caused by space objects'; more specifically 'damage caused by GNSS satellites'.[120] The term 'caused by' is actually greatly favoured by the international community since it could resolve the vexing question of causation so as to:

> allow for different tests of remoteness and causality which may be appropriate for different obligations or in different contexts, having regard to the interest sought to be protected by the relevant primary rule.[121]

This means that causation, in the context of GNSS damage, is open to being examined and tested on the basis of discretion and under each theory of causation in international law, which mainly refers to the criterion of 'directness',[122] 'foreseeability'[123] or 'proximity'.[124,125]

For the criterion of 'directness', the international community, including the United Nations Compensation Commission (UNCC),[126] has started to abandon knowledge accumulated from old arbitral decisions[127] which qualify damage not immediately caused by the wrongful act as 'indirect', and exclude this kind of damage for

117. I.H.Ph. Diederiks-Verschoor & Pablo Mendes de Leon, *An Introduction to Air Law* (Kluwer Law International, 2012), at 302; Elmar Giemulla & Ronald Schmid (Eds), *Montreal Convention* (Kluwer Law International, 2010), at Article 16-4.
118. Matte, *supra* note 110.
119. *See* Article XIV, the Liability Convention.
120. As discussed above in §3.03[B], there are few possibilities to interpret 'GNSS signal' as 'space object', so here the author will not discuss the causation between 'damage' and 'GNSS signal' under the phrase 'damage caused by space objects'.
121. *See* Article 31(1) of the Responsibility of States for Internationally Wrongful Acts; James Crawford, *Articles on Responsibility of States for Internationally Wrongful Acts*, http://legal.un.org/avl/pdf/ha/rsiwa/rsiwa_e.pdf, last accessed 16 August 2017.
122. *See* para. 16 of the Security Council resolution 687 (1991).
123. *See Portuguese Colonies case (Naulilaa incident)*, in United Nations, Reports of International Arbitral Awards: Vol. II (Sales No. 1949.V.1), at 1031.
124. *See* William Lloyd Prosser, *Selected Topics on the Law of Torts: Five Lectures Delivered at the University of Michigan* (William S. Hein, 1982), at 191.
125. United Nations, *Yearbook of the International Law Commission 2000: Volume II Part One* (United Nations, 2009), at 18.
126. The UNCC was created in 1991 as a subsidiary organ of the United Nations Security Council under Security Council resolution 687 (1991) to process claims and pay compensation for losses and damages suffered as a *direct* result of Iraq's unlawful invasion and occupation of Kuwait in 1990-91. For more information, please *see* http://www.uncc.ch/, last accessed 16 August 2017.
127. *See* United Nations, *Yearbook of the International Law Commission 1989: Volume II Part One* (United Nations, 1992), at 12 (note 63).

compensation.[128] Contrastingly, it has started to hold that 'directness' only focuses on the presence of a clear and an unbroken causal link between cause and effect.[129] Also, the Mixed Claims Commission (US and Germany)[130] insisted that:

> it matters not how many links there may be in the chain of causation [...], provided there is no break in the chain [...].[131]

In more academic language, the author enunciates that as long as the damage can be clearly and unmistakably traced back, link by link, to the act as the exclusive cause through a connected, though not necessarily direct, chain of events, the damage must be compensated.[132] This could be better understood with the following conclusion, after checking the cases containing the discussion of direct or indirect damages (*see* §3.03[C]):

> It is only true to say that in the majority of cases, in which the epithets 'direct' and 'indirect' are applied to describe the consequences of an unlawful act, they are in fact being used synonymously with 'proximate' and 'remote'.[133]

What the author could read from this conclusion was that the usage of 'direct' or 'indirect' has nothing to do with the criterion of 'directness', which means that the causal link is unbroken, but the remoteness of damage, i.e., 'proximate' or 'remote'.

For the criterion of 'foreseeability', in tort law it is generally required that the *existence* or *type* of damage must be reasonably foreseeable by a *reasonable person* at or before the time the accident occurred, and it does not matter whether the liable person actually expected that damage or not and whether the *extent* of that damage has been foreseen.[134]

The notion of 'proximity' or 'proximate cause' does not have a generally accepted meaning in practice,[135] and its definition is still in progress with too much disagreement among courts and scholars, even though this notion is one of ancient 'vintage' in legal history.[136] Therefore, judges in the court often instead to determine that whether the

---

128. Damage not immediately related to the wrongful act such as loss of earnings or profits has been clearly stated to be compensated by UNCC. *See* paras 5 and 20 of the Governing Council Decision no. 7, S/AC.26/1991/7/Rev.1, 17 March 1992.
129. Marco Frigessi di Rattalma & Tullio Treves (Eds), *The United Nations Compensation Commission: A Handbook* (Kluwer Law International, 1999), at 21.
130. The Mixed Claims Commission (United States and Germany) was set up to deal with the compensation of the US nationals for damage caused in the Lusitania disaster from the German Government, under the Treaty of Berlin, signed 25 August 1921.
131. United Nations, *Report of International Arbitral Awards: Mixed Claims Commission* (United States and Germany) (1 November 1923-30 October 1939): Volume VII (United Nations, 2006), at 29.
132. Clyde Eagleton, *The Responsibility of States in International Law* (Kraus Reprint, 1970), at 202.
133. Bin Cheng, *General Principles of Law as Applied by International Courts and Tribunals* (Cambridge University Press, 1987), at 243.
134. Helen Gubby, *English Legal Terminology* (Eleven International Publishing, 2016), at 133.
135. *See* William Lloyd Prosser, *Prosser and Keeton on the Law of Torts* (West Publishing Company, 1984), at 263-280.
136. William C. Bryson, *Cause and Consequence in the Law*, in Rom Harre & Fathali M. Moghaddam (eds), Questioning Causality: Scientific Explorations of Cause and Consequence across Social Contexts (ABC-CLIO, 2016), at 331.

damage is 'not proximate' or 'too remote'.[137] It should be noted here that the criterion of 'proximity' itself does not exclude all 'remote' causes, but only those which are 'too remote'.

Actually, neither international law, nor national law shows a general standard or theory for the matter of causation, and there is no clear line to make an exact judgment on the notions of 'directness', 'foreseeability', and 'proximity' respectively. This fact urges us not to focus on one criterion mechanically, but to remain flexible as long as *the principles of justice and equity*, which are also stipulated in Article XII of the Liability Convention, are duly observed. Therefore, the author holds that while the criterion of 'directness' intends to establish a factual causation, the criteria of 'foreseeability' and 'proximity' will transfer that factual causation to a legal one albeit with some limitations; this so as to avoid an infinity of possible parties liable for even minor acts of negligence, and to restrict it from going too far beyond what the generally shared sense of justice would support.[138]

Based on the analysis above, we could now try to test causation in the hypothetical case mentioned above (*see* §3.03[C]). In this hypothetical case, the three possible causes of GNSS malfunctioning – which, notably, the author believes cover most cases concerning GNSS civil liability – can be summarised as follows.

(i) Collision with Space Object (A).

In this case, the damage is actually caused by the collision between Space Object (A) and GNSS satellites. The causal link between Space Object (A)/GNSS satellites and damage must be established to claim compensation. While determining whose fault it is that caused the collision is critical for the identification of the liable party and the division of compensation in outer space,[139] it only makes a small difference to the causation test: if the collision were caused by Space Object (A), then the causal link would be illustrated as in Figure 3.2-A; if the collision were caused by GNSS satellites, the causal link would be illustrated as in Figure 3.2-B; and if the collision were caused jointly by Space Object (A) and GNSS satellites, the causal link would be illustrated as in Figure 3.2-C.

137. Herbert Lionel Adolphus Hart & Tony Honore, *Causation in the Law* (Oxford University Press, 1985), at lii.
138. Bryson, *supra* note 136, at 330.
139. Articles III & IV of the Liability Convention.

Figure 3.2 Causation (I)

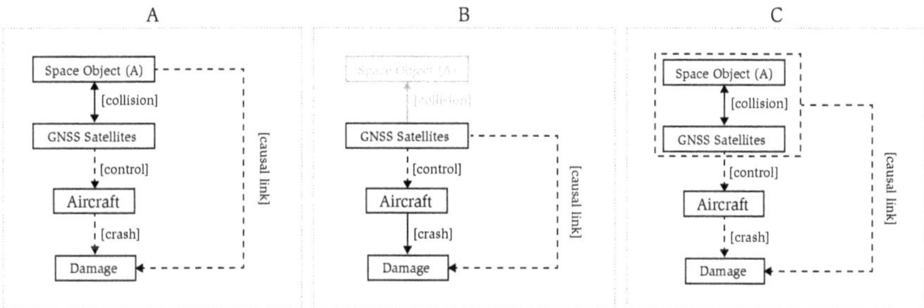

Compared with the typical case in space law, Figure 3.2 does not show any physical connection between the space object in question and damage, yet the causal link is unbroken and this complies with the criterion of 'directness'. If there is no other factor intervening in this causal link, a reasonable person would be aware, especially after the accident of the 'Iridium 33 and Cosmos 2251 Collision' which indicates the possibility of an interruption in communication service,[140] that the collision between a space object and GNSS satellites may interrupt GNSS signals or services.[141] This means that the criterion of 'foreseeability' also fits here. The only thing that needs to be further discussed is whether the causal link is too remote or not under the criterion of 'proximity'. As there is no clear standard for the notion of 'proximity', we have to make a weighing of interests, through the principles of justice and equity, between victims and potential liable parties.

It is clear that the Liability Convention favours *third* parties (*see* (ii) of §3.04) who are not involved in highly dangerous space activities,[142] and requires the liable party to provide compensation to the extent of placing the one being compensated in the situation that one would be in had the damage not occurred.[143] This notwithstanding, it would not be fair to hold the party liable for any consequence which is not very closely related to the starting point of the causal link. Bearing the above victim-oriented nature in mind, the author however believes that, since all the causal links in Figure 3.2 are simple, proximate, and not too remote, it is fair enough, in the sense of joint liability in outer space as shown by Figure 3.2-C, to hold the party who or whose fault caused that collision to make prompt and full compensation to the victims in this case.

140. *See* Iridium Satellite LLC, *Update on Iridium Satellite Constellation*, http://investor.iridium.com/releasedetail.cfm?ReleaseID=429190, last accessed 17 August 2017.
141. The space segment is a constellation of more than twenty satellites, and the collision or malfunction of a few satellites may not interrupt the GNSS service as a whole but it is still not impossible. The lack of enough satellites that function well at least makes a difference to the performance of GNSS signals, which may cause an air accident.
142. Marietta Benko, Kai-Uwe Schrogl, Denise Digrell & Esther Jolley, *Space Law: Current Problems and Perspectives for Future Regulation* (Eleven International Publishing, 2005), at 92.
143. Article XII of the Liability Convention.

(ii) Radio interference from Space Object (A).

In this case, the damage is caused by the radio interference from Space Object (A), and what is required for the claim is the causal link between Space Object (A) and damage, as shown in Figure 3.3-A.[144] The structure and remoteness of a causal link in this case is similar to the one in Figure 3.2-A, and the only big difference is between *collision* and *interference*, i.e., *physical connection* and *remote effect*. However, the matter of causation never requires a 'physical' link as an essential element. For example, in the 'Cosmos 954 Claim', the nuclear damage was not caused by a direct hit and connection, but by radiation contamination which was accepted as the proximate cause of harm,[145] while the compensation was finally granted by the USSR *ex gratia* without normative content despite Canada's claim based on the Liability Convention.[146] It seems too narrow, to be fair enough, to restrict the notion of damage to the damage caused exclusively by direct contact, and this argument is also shared by the theory of general law where air law is also included.[147] Actually, the key point is not whether the damage is suffered through physical impact with a space object, or whether it results from biological, chemical or radiological contamination emanating from a space object.[148]

Therefore, the author believes that if the causal link could be established for the damage caused by the *collision* between Space Object (A) and GNSS satellites (*see* (i) of §3.03[D]), there is no reason to deny the causal link for the damage caused by *radio interference* from Space Object (A), in particular considering that the damage at the end of that causal link complies with the consequence referred to in the Liability Convention.[149]

Figure 3.3 Causation (II)

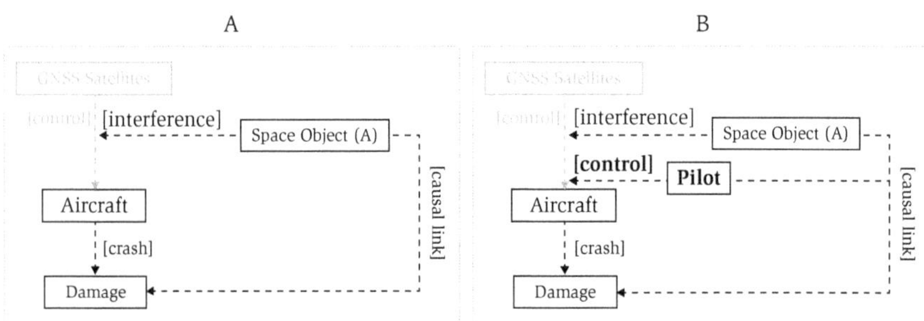

144. The Figure 3.3-B and Figure 3.4-B will be discussed with Figure 3.7 in §3.04, below.
145. Christol, *supra* note 48, at 359.
146. Alexander F. Cohen, *Cosmos 954 and the International Law of Satellite Accidents*, 10(1) Yale Journal of International Law 1984, at 87; Edward G. Lee & David W. Sproule, *Liability for Damage Caused by Space Debris: The Cosmos 954 Claim*, 26 Canadian Yearbook of International Law 1988, at 276.
147. *See* Diederiks-Verschoor & Mendes de Leon, *supra* note 117, at 302.
148. Foster, *supra* note 45, at 155.
149. Kayser, *supra* note 96, at 48.

(iii) Malfunction of GNSS itself.

In this case, the damage is caused by the GNSS itself, including its satellites,[150] where the causal link between GNSS satellites and damage has to be established for a relevant claim for compensation. Admittedly, the case of this model already existed,[151] and the author believes that it would continue to be the most possible way to happen in practice, compared with *collision* and *radio interference* in outer space. Compared with the above two models, the difference in this case is that no other space object but GNSS satellites could be blamed for the damage, and this makes the causal link (*see* Figure 3.4-A) even simpler and less remote, which is a good point for the victim.

Figure 3.4 Causation (III)

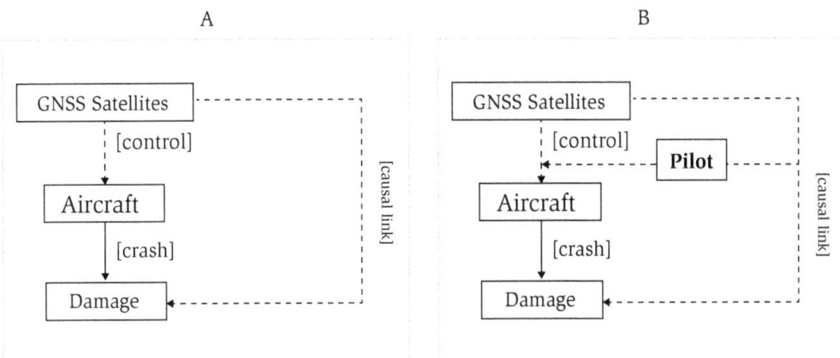

It is clear that there is no physical connection between space objects (GNSS satellites) and damage, but as discussed above, it does not affect the establishment of causal link. To support the author's argument, an analogy of the causal link between a case of damage caused by GNSS and a case by ATC could be made with reference to the well-known '2002 Überlingen mid-air collision', in which case the main cause was attributed to the ATC service provider.[152] The main causal link in the '2002 Überlingen mid-air collision' could be simply illustrated as in Figure 3.5-A.

150. One may argue in this case that the problem may arise from the ground control segment crashing rather than the failure of GNSS satellites themselves, but this does not affect the civil liability issues of GNSS satellites under international space law, as all users get (defective) signals from those satellites rather than ground transmitters. However, defective signals solely from a GBAS do not in any way apply to the outer space treaties (*see* (v) of §3.04), but it may be involved in legal disputes as the operator or provider has to prove its innocence, which is quite difficult.

151. *See Glonass Failure Caused by Faulty Software*, http://www.gpsdaily.com/reports/Glonass_Failure_Caused_by_Faulty_Software_999.html, last accessed 21 August 2017.

152. *See* German Federal Bureau of Aircraft Accidents Investigation, *Investigation Report*, AX001-1-2/02, May 2004, http://cfapp.icao.int/fsix/sr/reports/02001351_final_report_01.pdf, last accessed 21 August 2017.

Figure 3.5 Causation (IV)

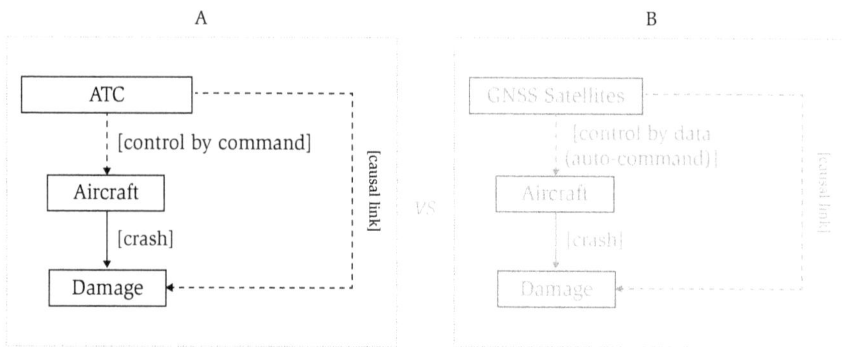

It is quite apparent that the structure and remoteness of the causal link in Figure 3.5-A is almost the same as the one in Figure 3.5-B (Figure 3.4-A), and the only difference lies in the fact that, while ATC controls an aircraft by the 'command' to the 'pilot-in-command', GNSS satellites control an aircraft by the 'signal' or 'data' to the 'GLS' which could be regarded as an 'auto-pilot'. Therefore, if the causation in Figure 3.5-A could be widely accepted by air law practitioners,[153] then there will be no reasonable excuse for space law experts to deny the one in Figure 3.5-B (Figure 3.4-A).

[E] Brief Conclusion

Based on the analysis above, it could be briefly concluded that GNSS damage may apply the Liability Convention, as well as its legal basis that is Article VII of the Outer Space Treaty, as long as the causal link between that damage and a GNSS satellite – not GNSS signal – is unbroken, logical, foreseeable, not too remote and, finally, accepted in diplomatic negotiations or by the Claim Commission on a case-by-case basis. Technically speaking, the author also admits that it will be quite challenging in practice to prove the matter of causation in the case concerning GNSS damage, but it is still not impossible with the help of, for example, technical experts.

In addition, the term 'space object' is of both a material and a physical property, and therefore a GNSS signal is excluded. The discussion on whether GNSS damage qualifies as indirect damage, and whether the Liability Convention covers indirect damage is irrelevant to the applicability of GNSS damage to outer space treaties.

153. The compensation issues between Bashkirian Airlines (whose aircraft crashed and who paid compensation to most victims in the crash according to air law) and the Federal Republic of Germany (which transferred its part of sovereignty in terms of ATC service to Skyguide and which is legally responsible for the ATC liability) reached an out of court settlement around 2013. *See* Francis Schubert, *The Liability of Air Traffic Control Agencies – The Ueberlingen Midair Collision Case Study*, Presentation to the Institute of Air and Space Law, McGill University, October 2014, at 51.

§3.04 REALISTIC: A LESS FEASIBLE MECHANISM FOR GNSS CIVIL LIABILITY

Arising from the ultra-hazardous nature of space activities, international space law intends to protect third-party victims mostly by a victim-/claimant-oriented liability framework such as a strict liability system, and an easily identifiable liable party. Unfortunately, certain gaps always exist between the ideal and the real. For one thing, we have to admit that, for the delegations of outer space negotiations and drafting of relevant treaties, which were mostly matters of thirty years ago, GNSS legal issues were at least, if not unpredicted, not obvious concerns.[154] For the other thing, what an expected regime of GNSS civil liability requires does not lie only in the fair, prompt, and adequate compensation for all potential victims, but also in the weighing of interests between users and providers as well as GNSS sustainable development from a long-term perspective. In this sense, the author holds that current international space law is neither an adequate, nor a fair mechanism for GNSS civil liability, despite the author having made herculean efforts as above to find a possible way to apply GNSS damage to Article VII of the Outer Space Treaty and the Liability Convention. The main reasons lie in the following aspects.

(i) Imperfect relationship between Article VII of the Outer Space Treaty and the Liability Convention.

It is generally accepted that the civil liability regime under international space law is jointly composed of Article VII of the Outer Space Treaty and the Liability Convention; the former being the legislative basis, the latter being the implementing rules thereof.[155] The author does not intend to make any negative comments on the status of the Outer Space Treaty and the Liability Convention,[156] although it seems clear that there are a number of States that are parties to the one but not the other.[157] This fact matters to GNSS civil liability, bearing in mind that the opinion to separate the discussion of the applicability of the Outer Space Treaty and the Liability Convention in the case of GNSS

154. Jennifer Ann Urban, *Soft Law: The Key to Security in a Globalized Outer Space*, 43 Transport Law Journal 2016, at 45; Carl Q. Christol, *Space Law: Past, Present, and Future* (Kluwer Law and Taxation Publishers, 1991), at 223; Foster, *supra* note 45, at 158.

155. Hurwitz, *supra* note 1, at 9; Hobe, Schmidt-Tedd & Schrogl, *supra* note 26, at 136; Hamilton DeSaussure & Peter P.C. Haanappel, *A Unified Multinational Approach to the Application of Tort and Contract Principles to Outer Space*, 6 Syracuse Journal of International Law and Commerce 1978, at 2; Elie Robert Camille van Bogaert, *Aspects of Space Law* (Kluwer Law and Taxation Publishers, 1986), at 162.

156. As of 1 January 2018, there are 107 and 95 Member States of the Outer Space Treaty and the Liability Convention respectively. Although not all States own a space industry and only a few States qualify as GNSS farmers, each State and its nationals is part of GNSS users group and potential victims suffering from GNSS damage. Even though it is our best wish to include the actors as much as possible in an international legal framework, we have to admit that we may not expect too much as both Outer Space Treaty and the Liability Convention have already made great achievements in terms of the number of Member States, especially compared with the Moon Agreement. *See* UNOOSA, *Status of International Agreements Relating to Activities in Outer Space as at 1 January 2018*, A/AC.105/C.2/2018/CRP.3, 9 April 2018.

157. *Ibid.*; Hurwitz, *supra* note 1, at 10.

damage already exists.[158] Article VII of the Outer Space Treaty is too general to define the nature, and to detail rules, of international liability itself,[159] but this generality may be used as an excuse to escape from the regulation of the Liability Convention. For a Member State of the Outer Space Treaty – or if we accept the status of customary international law, for any State – which is not a Member State of the Liability Convention, specific terms of a victim-oriented nature such as the absolute liability regime under the Liability Convention may not be binding,[160] and this would decrease the victim's wish to apply international law as a whole in the case of GNSS damage.

(ii) Inadequate scope of application.

If we recall the constitution of GNSS (*see* §1.02[B] and Figure 1.2), both SBAS and GBAS are parts of GNSS. It seems that there is no question about the applicability of international space law to SBAS based on the nature of space systems. Yet, neither the Outer Space Treaty nor the Liability Convention applies to the damage caused solely by a GBAS, since it itself is not a space system, and the augmented signal is transmitted from ground facilities. In addition, the Liability Convention uses a *third-party* liability system.[161] The Liability Convention also does not apply to nationals of the liable launching State of GNSS satellites, and it excludes foreign nationals who are involved in the launching and operation of GNSS satellites.[162] This situation would not benefit the case of GNSS. The reason is that the scope of the launching State may be so comprehensive by virtue of the term 'procures',[163] which could refer to the State itself or its nationals who provide the financial capital for the launch,[164] that it would leave too much room for different interpretations.[165] For example, not only have EU Member States and States such as the Swiss Confederation, Ukraine, Norway, and China participated in both the development and operation of the EU's Galileo particularly in the form of funding,[166] but in the future, the scope of cooperation partners is likely to be enlarged. Involvement in Galileo is so wide-ranging that the scope of application of the Liability Convention will be constricted too much.[167] If we expect a civil GNSS

158. *See* Perez, *supra* note 42, at 45.
159. John M. Kelson, *State Responsibility and the Abnormally Dangerous Activity*, 13 Harvard International Law Journal 1972, at 215.
160. Hurwitz, *supra* note 1, at 10; Ram. S. Jakhu, Diane Howard & Andrea J. Harrington, *Legal Aspects of Solar Power Satellites*, in Leslie I. Tennen (eds), Private Law, Public Law, Metalaw and Public Policy in Space (Springer International Publishing Switzerland, 2016), at 29.
161. Kerrest & Thro, *supra* note 5, at 64; Frans von der Dunk, *The International Legal Framework for European Activities on Board the ISS*, in Frans von der Dunk & Marcel M. T. A. Brus, The International Space Station: Commercial Utilisation from a European Legal Perspective (Martinus Nijhoff Publishers, 2006), at 17.
162. Article VIII of the Liability Convention.
163. Article VII of the Outer Space Treaty; Article I of the Liability Convention.
164. Hurwitz, *supra* note 1, at 22.
165. Hobe, Schmidt-Tedd & Schrogl, *supra* note 45, at 114.
166. EC, *Reference Documents*, http://ec.europa.eu/growth/sectors/space/galileo/documents_en, last accessed 11 September 2017.
167. The EU has neither signed nor ratified outer space treaties as an international organisation, but most Member States of the EU and the ESA (co-partner in the Galileo programme) are under the governance of outer space treaties. Also, the ESA itself made a declaration of acceptance of the Rescue Agreement, the Liability Convention and the Registration Convention. *See* respectively:

whose programme partner may come from every corner of the globe, where certain GNSS users may also be included, the above negative effect would be exacerbated.

(iii) State-versus-State liability system.

For ultra-hazardous activities of a global nature such as those about nuclear, oil pollution, and outer space, the treaty practice prefers a *State liability* regime,[168] where:

> States have been held liable for injuries caused to other States and their nationals as a result of activities occurring within their territorial jurisdiction or under their control.[169]

Considering the fact that a *State*, compared with a *private* legal entity, is much more identifiable and with abundant capital, a State liability system is seemingly one of the best solutions to protect victims in a disaster caused by space objects as well.[170] Although it is without question that Article VII of the Outer Space Treaty, and especially the Liability Convention, is based on the premise of State liability, this is rather unique.[171] Claims must be brought by a 'State' against another 'State',[172] and the Liability Convention does not grant a private party the right to present a claim for compensation so much so that the individual must petition his/her government to make that claim.[173] The problem however is that the States from both sides may not

Andreas Loukakis, *EU as Owner of Galileo Satellites: Consequences for Registration and Liability*, in Mahulena Hofmann & Andreas Loukakis (eds), Ownership of Satellites: 4th Luxembourg Workshop on Space and Satellite Communication Law (Nomos, 2017), at 131; ESA, *Declaration of Acceptance of the Astronauts Agreement, the Liability Convention and the Registration Convention*, adopted by the ESA Council on 12 December 1978, deposited on 2 January 1979.

168. The term 'State liability' in this book is different with the one established by the European Court of Justice. For the latter, please *see Francovich and Bonifaci v. Italy*, Case C-6/90 and C-9/90, [1991], E.C.R. I-5357; Sheila Bone (Eds), *Osborn's Concise Law Dictionary* (Sweet & Maxwell Limited, 2001), at 360.
169. UN Secretariat, *Survey of Liability Regimes Relevant to the Topic of International Liability for Injurious Consequences Arising Out of Acts Not Prohibited by International Law*, Document A/CN.4/543, at 154-156.
170. Christian Brünner (Eds), *National Space Law: Development in Europe-Challenges for Small Countries* (Böhlau Verlag Wien, 2008), at 68.
171. Frans von der Dunk, *Passing the Buck to Rogers: International Liability Issues in Private Spaceflight*, 86 Nebraska Law Review 2007, at 410.
172. Article VII of the Outer Space Treaty is too general to understand whether its nature is a State-vs-State liability system. There is one opinion in the Cologne Commentary on Space Law (Volume I) which indicates that the Outer Space Treaty may be used for a claim by an individual victim before a national court. However, the author agrees with Professor Paul B. Larsen's opinion that *"The 1967 Outer Space Treaty is not intended to cover the issue of the jurisdiction of national courts."* Also, as the Liability Convention qualifies as implementing rules based on regulation based on Article VII of the Outer Space Treaty, we will not separate them too far in such terms as the status of international customary law and the nature of State-vs-State liability system, otherwise much more legal uncertainty will be created. *See* respectively: Hobe, Schmidt-Tedd & Schrogl, *supra* note 26, at 135; Paul B. Larsen, *UNIDROIT Space Protocol: Comments on the Relationship Between the Protocol and Existing International Space Law*, 44 Proceedings of the International Institute of Space Law 2001, at 191.
173. Rupert W. Anderson, *The Cosmic Compendium: Space Law* (Lulu.com, 2015), at 21; Luke Punnakanta, *Space Torts: Applying Nuisance and Negligence to Orbital Debris*, 86 Southern California Law Review 2012, at 174.

respectively be the actual liable party and the actual victim, and this leads to too many negative situations in practice.

First, taking so many political interests and deals between States into consideration, it is possible that the State, as is its right, decides not to present a claim for the injured party including individuals or private organisations,[174] in which case there is no other alternative way for the injured party in question but to submit a claim in the courts or administrative tribunals or agencies of a launching State.[175] By doing so the victim-oriented liability regime under the Liability Convention has no standing since that regime is only for State-versus-State claims, and the Liability Convention itself does not offer any juridical rules for a claim based on either international or domestic private law. Furthermore, the possibility for the victim to pursue a domestic remedy indicates in turn, to some extent, that the Liability Convention itself may not be adequate as the *legal* remedy, as opposed to *diplomatic* channels, for victims.

Second, at the first stage the Liability Convention only supports the State to raise a claim through traditional *diplomatic* negotiations, rather than *juridical* proceedings which have a more legally binding effect.[176] At the second stage, although a Claim Commission, also by *States*, could be established after diplomatic negotiations have failed, the decision of the Claim Commission would be only binding 'if the parties have so agreed',[177] and the implementation procedure of that decision remains open.

Third, States or international intergovernmental organisations, rather than their nationals who may be the actual liable party, are the only entities which can possibly incur international liability under the Liability Convention.[178]

Fourth, no clear procedure in the Liability Convention requests the State that successfully presents a claim to transfer the compensation funds to the actual victims, even though, while still not impossible, it is unlikely to happen.

Fifth, the Liability Convention's short limitation period of one year may also constitute an incentive to bring claims before national courts,[179] but again, the Liability Convention has no standing in a national court for private compensation.

It is true that the above aspects are general problems of the civil liability regime under the Liability Convention, but the risk of not presenting claims and transferring compensation as well as the uncertainty of traditional diplomatic channels can only be more obvious in the case of GNSS damage than in the case of the direct crash or collision of space objects. Victims in the context of GNSS have to go through a 'tedious' diplomatic process under the Liability Convention, as illustrated by Figure 3.6-A, which should have been very simple juridical proceedings against the liable party or appropriate State, as illustrated in Figure 3.6-B and Figure 3.6-C.

174. Hurwitz, *supra* note 1, at 49-50.
175. Article XI (2) of the Liability Convention.
176. Article IX of the Liability Convention; Milde, *supra* note 48.
177. Article XVIII (2) of the Liability Convention.
178. Frans von der Dunk, *Sovereignty Versus Space – Public Law and Private Launch in the Asian Context*, 5 Singapore Journal of International & Comparative Law 2001, at 38.
179. Chatzipanagiotis & Liperi, *supra* note 45, at 166.

Figure 3.6 Claim Procedure

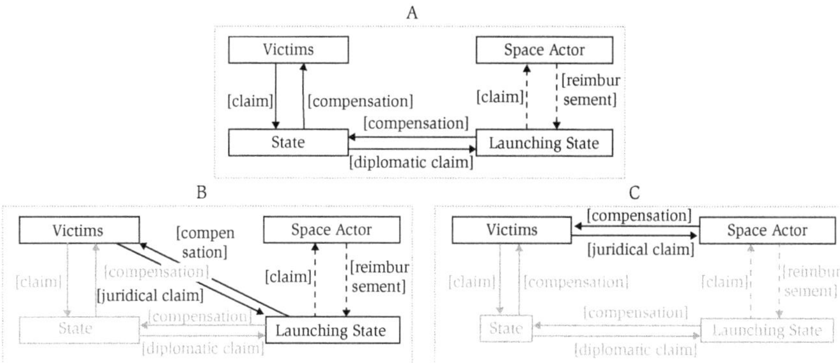

(iv) Serious legal uncertainty.

As discussed above, so far, no general standard or theory for the matter of causation exists to draw lines between two sides (yes and no) of the notions of 'directness', 'foreseeability'. and 'proximity'. The interpretation of these notions and terms appearing in outer space treaties will finally be the task of negotiating teams in diplomatic channels, or of the Claims Commission which can be established if those diplomatic actions to settle the claim fail.[180] On the one hand, the diplomatic channels themselves refer to much uncertainty for victims in the jungle of legal politics; on the other hand, it would not be surprising if the Claims Commission were to deny the proximity of those causal links outlined in Figure 3.2, Figure 3.3-A, and Figure 3.4-A as it is true that neither the collision between Space Object (A) and GNSS Satellites, nor the radio interference from Space Object (A) is the nearest cause of the damage, but rather it is the final aircraft crash. This conflict results from different understandings of remoteness of causal link in the context of GNSS cases, thus creating serious legal uncertainty; such a situation would be a nightmare both for victims and the potentially liable party.

In addition, the analysis in §3.03[D] is reasonable only if there are no external factors which are neither initiated by Space Object (A), nor by GNSS satellites that break the causal link. For example, if we say the aircraft control was taken over by the pilot in command after the alarm of either GLS or other on-board system, or if the pilot in command does not (duly) take action after those alarms, the causal links will be interrupted by the factor of the pilot in command, as illustrated in Figure 3.7, Figure 3.3-B, and Figure 3.4-B. In this sense, the 'directness' of causation does not exist anymore, and at least it will not be appropriate to claim absolute causation between the damage and GNSS satellites or Space Object (A). This would definitely challenge the applicability of the Liability Convention as well as the Outer Space Treaty in the case of GNSS damage.

180. Hurwitz, *supra* note 1, at 51.

Figure 3.7 Causation (V)

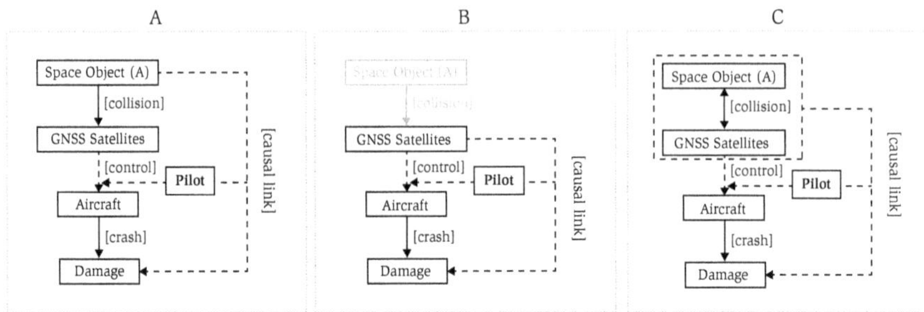

It should also be noted that neither Article VII of the Outer Space Treaty, nor the Liability Convention establishes a clear rule to calculate compensation; it would have been difficult for most States to support the Convention on the one hand,[181] and may lead to different standards for that calculation in the same or similar cases, thereby constituting another kind of legal uncertainty about the expected amount of compensation.

(v) Launching State versus Operating State.

According to the current civil liability regime under international space law, the identification of the liable party is focused on the launching activities.[182] However, some damage does occur long after those launching activities because of maloperation of in-orbit space objects, rather than the failure of the launching task before the operation period. Since launching activities do not continue for the life of the orbit, it would be unfair to make a launching State conceivably liable for the damage caused by a space object and its payload, particularly in the sense of launching a foreign rocket in its territory.[183] As only the party having effective control can limit the risk of an accident,[184] it would only make sense to make the operator or operating State, rather than the launching State, responsible and liable for the damage caused by the operation of a space object during its life in orbit. This is similar to the regime under Article II of the Convention on the Liability of Operators of Nuclear Ships (Brussels, 1962) where it is the operators who are supposed to be liable for the damage, rather than the manufacturer of nuclear ships.

In the context of GNSS, the situation above is particularly true as the damage is actually caused by space object operations and not the space object itself. On the one hand, when the malfunction of a GNSS satellite results from the crash of *ground* facilities or defective uplink data from those facilities, it clearly has nothing to do with the *launching* State which is too far from the control over those *ground* facilities in

181. *Ibid.*, at 53.
182. Article VII of the Outer Space Treaty; Article I (c), Article II and Article III of the Liability Convention.
183. Hobe, Schmidt-Tedd & Schrogl, *supra* note 26, at 137.
184. Kerrest & Thro, *supra* note 5, at 61.

some cases.[185] Hence, it is very inappropriate to make the launching State liable for that damage. On the other hand, damage caused by GNSS is mostly during the operation of GNSS satellites, rather than during the period of launching. If we recall the hypothetical case model (*see* §3.03[C]), the damage caused by GNSS is mainly incurred from the malfunction of GNSS itself during the operation process (*see* §1.03 and (iii) of §3.03[D]), even though, in the case of a collision between Space Object (A) and GNSS satellites and radio interference from Space Object (A) in outer space, the damage is beyond the control of the *launching* States, but is within the control of the operating States of Space Object (A) or GNSS satellites.

Both the Outer Space Treaty and the Liability Convention attribute liability to the launching State. However, the author argues that a launching State does not play a critical role in the malfunction of GNSS, and it seems the operating State is the more obvious liable party. GNSS users may recognise the provider of GNSS signals and they usually do not care who launched those GNSS satellites. Moreover, the launching State does not necessarily undertake *responsibility*, as provided by Article VI of the Outer Space Treaty, to supervise the operation of GNSS including its satellites and ground facilities. In the development of both new technologies such as GNSS, which addresses operation, and new practices such as ownership transfer of in-orbit space objects,[186] the legal ground of the launching State liability system under international space law seems increasingly outdated.

(vi) Against GNSS sustainable development.

GNSS is often viewed as a key enabler to support sustainable development of the world,[187] even though itself and the relevant industry also need sustainable development,[188] perhaps with the support from the national 'GNSS Sustainable Development Authority' (*see* (v) of §1.04[B][2]). The core nature of sustainable development is to ensure a better future for human beings, but it is read from different perspectives such as national governing philosophy,[189] environment protection,[190] and prosperity for

185. It is true that certain launching State is the State which operates the ground facilities of GNSS but that is not true for all.
186. Armmel Kerrest, *Legal Aspects of Transfer of Ownership and Transfer of Activities*, in Mahulena Hofmann & Andreas Loukakis (eds), Ownership of Satellites: 4th Luxembourg Workshop on Space and Satellite Communication Law (Nomos, 2017), at 77.
187. United Nations, *A Global Geodetic Reference Frame for Sustainable Development*, A/RES/69/266, 26 February 2015, at 2; United Nations, *International Cooperation in Peaceful Uses of Outer Space*, A/RES/61/111, 15 January 2007, at 8.
188. China Satellite Navigation Office, *Report on the Development of BeiDou Navigation Satellite System (Version 2.2)*, December 2013; National Medium and Long-term Planning for Satellite Navigation Industry Development, No. 97/2013 of the General Office of the State Council of the People's Republic of China.
189. The Theory Bureau of the Publicity Department of the Central Committee of the Communist Party of China, *Introduction to the Scientific Outlook on Development* (Central Compilation & Translation Press, 2006), at 40.
190. World Commission on Environment and Development, *Our Common Future* (WECD, 1987), at ES-7.

all.[191] Specific to the context of GNSS, the principle of sustainable development could be expressed simply as long-term sustainability of GNSS. Although it is our priority to protect victims' interests because of their innocence, it should not be an excuse to opt for a tragedy which 'kills the goose that lays the golden egg'.[192]

The concept of sustainable development does imply limits, although it does not mean absolute limits.[193] The competing interests between potentially liable party and victim should be duly coordinated in the context of GNSS civil liability for the long-term development of GNSS. In spite of its victim-oriented nature, the preamble of the Liability Convention has already indicated that compensation should not only be *full*, but also *equitable*.[194] Nevertheless, neither Article VII of the Outer Space Treaty, nor the Liability Convention has set limits in terms of compensation; the issue with this is that different limits hold for the affordability of GNSS providers, and this may cripple GNSS industries. An ideal GNSS civil liability regime should ensure sustainable benefits for all, including the future generations, by such regimes as the two-tier liability in air law system, limitation of liability, compulsory insurance or compensation fund, while guaranteeing adequate compensation for victims. From this perspective, the absolute liability regime without limitation adopted by international space law is not a practical choice in the context of GNSS civil liability regime, unless this unlimited liability is insured by valiant insurers which is unlikely to happen.

§3.05 CONCLUDING REMARKS

International space law is practising in the gap between the ideal and the real in terms of GNSS civil liability. While a possible way to apply Article VII of the Outer Space Treaty and the Liability Convention to the case concerning damage caused by GNSS satellites does exist, it is however less feasible to guarantee prompt, adequate, and equitable compensation to victims of damage caused by GNSS.

It is beyond question that the development and operation of GNSS qualify as space activities, meaning that international space law applies. Yet, the author believes that international space law is quite reticent to broadly interpret GNSS *signals* under the term 'space object'. The questions of (i) whether indirect damage is covered by Article VII of the Outer Space Treaty and the Liability Convention, and (ii) whether GNSS damage qualifies as indirect damage have seriously troubled the space community for quite a while, even though the answers to those two questions are not a case in point to apply civil liability regime under the Outer Space Treaty and the Liability Convention. The real crux lies in the establishment of causation, on a case-by-case basis, between damage and space objects where GNSS satellites are included, with

191. United Nations, *Sustainable Development Goals*, http://www.un.org/sustainabledevelopment/sustainable-development-goals/, last accessed 13 September 2017.
192. For more information on this proverb, please *see* Aesop, translated by R. Worthington, *Aesop's Fables* (The Floating Press, 2008), at 98.
193. World Commission on Environment and Development, *supra* note 190, at 2-1.
194. Hurwitz, *supra* note 1, at 9.

reference to the criterion of 'directness', 'foreseeability' and/or 'proximity' under the sense of general international law.

The author admits that, this notwithstanding, when both the Outer Space Treaty and the Liability Convention were formulated, it was far too early to handle or even foresee such 'new' (comparatively speaking) space technology concerns as GNSS civil liability. Current international space law provides neither an adequate, nor a fair mechanism for GNSS civil liability. This is particularly so when considering its unique nature of State-versus-State liability and the launching-State-focused liability system without limitation, as well as legal uncertainty arising from different understandings of the matter of causation especially in the case of the intervention of external factors.

CHAPTER 4

International Air Law for GNSS Civil Liability: An Alternative Solution?

§4.01 INTRODUCTION

International space law traditionally regulates space activities of a global nature, whereas international air law governs the utilisation of space technologies such as satellite communication and navigation for international civil aviation.[1] Despite the space-based nature of GNSS, current international space law provides neither an adequate, nor a fair mechanism regarding GNSS civil liability (*see* Chapter 3). Considering the long period of discussion about a legal framework of GNSS and serious concerns about the issue of civil liability in the international aviation community represented by ICAO (*see* §4.03), attention to a transition from international space law to international air law is now necessary.

This chapter aims to examine whether, and if so, to what extent current international air law could be an alternative solution for GNSS civil liability. For that purpose and as an essential background, the connection between GNSS and international air law is first established (*see* §4.02). This chapter then proceeds from international public air law to international private air law; although the Chicago Convention does not regulate civil liability, it does lay down the basic rules of State responsibility on air navigation facilities where GNSS plays a fundamental role in international civil aviation.

Therefore, this chapter examines whether the present regime of ATC civil liability could cover the issue of GNSS civil liability (*see* §4.04). Subsequently this chapter review the applicability of international treaties, concluded from Warsaw to Montreal via Rome, regarding air carriers' civil liability for damage caused by GNSS malfunction

1. *Cf.* Boakye Danquah Kofi Henaku, *The ICAO CNS/ATM System: New King, New Law?*, XIX(3) Air & Space Law 1994, at 151.

or defective PNT signals through the operation of aircraft (*see* §4.05). The chapter closes with concluding remarks (*see* §4.06).

§4.02 GNSS UNDER INTERNATIONAL AIR LAW

[A] Legal Implications of GNSS for International Civil Aviation

Nowadays, GNSS benefits society as a whole from both an economic and a safety perspective, and by far, it is commonly used for the safety of navigation in different modes of transportation, ranging from deep sea to outer space.[2] Aviation is one of the sectors that benefits most from the use of GNSS.[3] Even though aviation experts once expected to implement an aeronautical satellite system exclusively for civil aviation, the lack of financial and procedural feasibility eventually defeated them.[4] Accordingly, the aviation community has no choice but to share GNSS with other sectors such as maritime transport. Due to this fact, for one thing, there is a need for the civil aviation community to closely coordinate with other relevant sectors and organisations in the use of GNSS, such as the IMO.[5] For another thing, it in turn demonstrates that civil aviation cannot be considered in isolation when addressing technical and legal issues of GNSS, including civil liability and cost-sharing mechanisms.[6]

As a result of the highly sensitive nature of safety, the international civil aviation community is seemingly the most active group which has been conducting continuing discussions about the legal framework of GNSS and presenting particular concerns about the issue of civil liability since the early stages of GNSS utilisation. Ocean liners nowadays have a considerable amount of freight on board, though mainly for *cargo* transport,[7] and the historical importance of such ocean liners for *passenger* travel has been challenged by the development of civil aviation.[8] In this regard, passenger safety and relevant legal issues are not addressed as frequently and urgently by the maritime community when compared with those in the context of civil aviation.[9] Besides, other

2. Peter J.G. Tenunissen & Oliver Montenbruck (Eds), *Springer Handbook of Global Navigation Satellite Systems* (Springer International Publishing, 2017), at 20.
3. Isabella Henrietta Philepina Diederiks-Verschoor & Vladimír Kopal, *An Introduction to Space Law* (Kluwer Law International, 2008), at 67.
4. Ruwantissa Abeyratne, *Frontiers of Aerospace Law* (Routledge, 2002), at 35.
5. Ludwig Weber, *The Global Navigation and Communications Satellite Systems and the Role of ICAO*, in ESA/ECSL, et al., Proceedings of the Third ECSL Colloquium – International Organisations and Space Law (European Space Agency, 1999), at 98.
6. Francis P. Schubert, *An International Convention on GNSS Liability: When Does Desirable Become Necessary*? XXIV Annals of Air and Space Law 245, 1999; ICAO, *GNSS – Cost Allocation*, https://www.icao.int/sustainability/Pages/eap-im-gnss-cost-allocation.aspx, last accessed 11 November 2017.
7. *See* Sushil K. Gupta, *The Routledge Companion to Production and Operations Management* (Taylor & Francis, 2017), at 559.
8. For example, as far back as 1958, airlines carried more passengers than ocean liners across the Atlantic for the first time. John Bowen, *Air Transport*, in Jean-Paul Rodrigue, Claude Comtois & Brian Slack, The Geography of Transport Systems (Taylor & Francis, 2016), at 148.
9. Although IMO issued its 'Maritime policy for a future Global Navigation Satellite System (GNSS)' in 1997 in which legal aspects such as principles on cost recovery and civil liability were slightly addressed (*see* §1.04[B][3]), but the author regrettably does not see any further development in

sectors – such as road transport – or non-professional areas in the use of GNSS – such as private car-driving or yachting – which lack specific international governmental or non-governmental organisations to mandate the collective action on such legal issue on GNSS civil liability are not as internationally and heavily regulated.[10]

Therefore, this section first examines the connection point between civil aviation and GNSS technology, and then reviews the initiatives conducted by ICAO for associated legal implications on GNSS. Furthermore, this section classifies international legal instruments which potentially cover the issue of GNSS civil liability as a basis for further study in the following sections.

[B] The Use of GNSS for International Civil Aviation

Technically speaking, GNSS is a space technology intending to offer seamless satellite navigation services through the capability of determining the real-time movement track parameters of an aircraft in the course of a flight.[11] To accommodate the need of aviation for very high-level performance parameters in terms of accuracy, integrity, continuity, and availability (*see* §1.02[D]), augmentation systems such as EGNOS and GPS-aided GEO Augmented Navigation (GAGAN) are required, in many cases, along with core GNSSs.[12] Aviators throughout the world use GNSS to increase both safety and efficiency of flight,[13] and this usually works through a certified GNSS receiver on board aircraft. Succinctly, the approaches that GNSS uses in its role to aviation users depend on the following two safety-critical applications:[14]

(i) Inputting PNT data directly to such decision-making system as *true* autopilot (autonomous) system on board.[15] GNSS, in this approach, works as one essential element of independent decision-makers without human pilots'

---

this aspect, apart from the updated recognition and acceptance of GNSS suitable for international use. For more information on the IMO's initiatives on GNSS, *see* Hiroyuki Yamada, *IMO and the GNSS: Navigating the Seas*, (September/October Inside GNSS 2017), at 40-44.

10. *Cf.* Frans von der Dunk, *Liability for Global Navigation Satellite Services: A Comparative Analysis of GPS and Galileo*, 30 Journal of Space Law 2004, at 130.
11. ICAO, *Global Navigation Satellite System (GNSS) Manual*, Second Edition-2013, Doc 9849, AN/457, at 1-1.
12. As the urgent need for high-quality navigation signals rises, certain GNSS users in aviation are even participating in the development of a SBAS. For example, the development of EGNOS is the result of a tripartite agreement between the ESA, the European Commission and EUROCONTROL in 1998; the Operator of EGNOS (ESSP) was founded in 2001 by seven European ANSPs; Indian GAGAN was developed by the ISRO, together with AAI. *See* respectively: GSA, *EGNOS Governance*, https://www.gsa.europa.eu/european-gnss/egnos/programme/governance, last accessed 10 November 2017; Aiylam Subramanian Ganeshan, et al., *India's Satellite-Based Augmentation System: GAGAN – Redefining Navigation over the Indian Region*, January/February InsideGNSS 2016, at 42.
13. *Aviation*, https://www.gps.gov/applications/aviation/, last accessed 10 November 2017.
14. In terms of criticality the GNSS applications can be classified in three different groups: (i) safety-critical applications; (ii) liability critical applications; (iii) non-Critical applications. For more information, *see* GMV, *Criticality of GNSS Applications*, http://www.navipedia.net/index.php/Criticality_of_GNSS_Applications, last accessed 11 November 2017.
15. FAA, *Advanced Avionics Handbook* (FAA, 2009), at 4-6.

interference. At this time, the said true autopilot may theoretically happen throughout all phases of flight of aircraft, especially for advanced drones.[16]

(ii) Providing three-dimensional position and velocity of aircraft to decision-makers, either pilots or air traffic controllers. The flight control in this approach involves human intervention, and GNSS is merely a complementary system providing navigation aid or guidance,[17] rather than a final decision-maker.[18]

As the UN specialised agency to, *inter alia*, "develop the principles and techniques of international air navigation,"[19] ICAO is responsible for developing the position and determining the requirements of all matters related to the use of space technology for international civil aviation.[20] The aforementioned utilisation of GNSS for air navigation purposes is included in this mandate.[21] In addition, considering the geographical limitation of terrestrial-based air navigation systems in the 1980s, ICAO recognised the need for the most cost-effective and efficient solution as an alternative to cater for evolving *global* air traffic demand.[22] Therefore, in 1983 the ICAO Council established a special committee entitled 'Future Air Navigation Systems (FANS) Committee' to study, identify, and assess the pros and cons of the use of new space technology such as GNSS for civil aviation purposes.[23] In 1988, after extensive study, the FANS Committee developed a groundbreaking new concept named the CNS/ATM systems.[24] This concept gained universal approval in 1991 at the tenth air navigation conference of ICAO (Montreal) with the backup of the International Air Transportation Association (IATA),[25] and in 1992 it was endorsed with two resolutions by the twenty-ninth Session of the ICAO Assembly.[26]

---

16. Autopilot integrated with GNSS is capable of providing control of the aircraft throughout each phase of flight, but in practice the pilot usually engages the autopilot between takeoff and landing. SKYbrary, *Autopilot*, https://www.skybrary.aero/index.php/Autopilot, last accessed 10 November 2017. Notably, a few automakers such as Tesla have already provided cars with full self-driving capacity based on Enhanced Autopilot to the public. *See* Tesla, *Autopilot*, https://www.tesla.com/autopilot, last accessed 11 November 2017.
17. GSA, *EGNOS: Making Landing Approaches More Precise and Efficient*, https://www.gsa.europa.eu/news/egnos-making-landing-approaches-more-precise-and-efficient, last accessed 12 November 2017.
18. From a legal point, the first approach addresses the decision-making role played by GNSS in the steering of an aircraft; the second approach underlines the supporting role played by GNSS during the control and guidance of aircraft. The above difference in the role of GNSS will make a large difference in the allocation of civil liability.
19. Article 44 of the Chicago Convention.
20. Weber, *supra* note 5, at 98.
21. *See* United Nations, *Report on the Second United Nations Conference on the Exploration and Peaceful Uses of Outer Space*, Vienna, 9-21 August 1982, A/CONF.101/10, at 100-101.
22. ICAO, *CAR/SAM Regional Plan for the Implementation of the CNS/ATM Systems: Document I*, September 1999, at 1-1.
23. Frans von der Dunk, *Navigating Safely Through the 21st Century: ICAO and the Use of GNSS in Civil Aviation*, 47 India Journal of International Law 2007, at 2.
24. ICAO, *Report of the Fourth Meeting of the Special Committee on Future Air Navigation Systems*, Doc. 9524, FANS/4, Rec. 2/1, 1988.
25. ICAO, *Report of the 10th Air Navigation Conference*, Doc 9583, 1991.
26. ICAO, *Appendix A: General Policy*, and *Appendix B: Harmonization of the Implementation of the ICAO CNS/ATM Systems*, to Consolidated statement of continuing ICAO policies and practices

The implementation of CNS/ATM systems started in 1989 when the 'Special Committee for the Monitoring and Coordination of Development and Transition Planning for the Future Air Navigation System (FANS Phase II)' was established. After four years' research, the FANS Phase II Committee developed the first plan of action.[27] In 1998, this plan of action was revised and retitled as the 'Global Air Navigation Plan for CNS/ATM Systems', and in it, legal and institutional elements, among others, began to be addressed. With an update in 2002,[28] this plan was again retitled, in 2007, the 'Global Air Navigation Plan' (GANP). The legal issues of CNS/ATM systems, mainly referring to GNSS, have been continuously addressed in the form of Appendix C of GANP up to current fifth edition issued in 2016 (*see* §4.03[B]).[29] Meanwhile, the regional plan and national plan for the implementation of CNS/ATM systems are parallel with the global plan. The regional and national approaches have been reflected in terms of proposed regional solutions for GNSS civil liability by the civil aviation community (*see* §4.03[B]).[30] The evolution of the concept of CNS/ATM systems and their implementation can perhaps be understood better with reference to the Annex.

However, regardless of the general acceptance of the concept of CNS/ATM systems and the various plans of action, the implementation was considered as probably the most complex and far-reaching project ever undertaken by ICAO in the history of aviation.[31] It is not easy to achieve a successful transition from the currently used air navigation system to a new system considering both the difficulty of moving such a large number of States and organisations forward towards new technology,[32] and the serious concerns, particularly addressed by the GNSS user States, about the lack of a legal framework for CNS/ATM systems (*see* §4.02[C]).

---

related to communications, navigation and surveillance/air traffic management (CNS/ATM) systems, Assembly Resolution A31-6, 1995.

27. ICAO, *Global Coordinated Plan for Transition to ICAO CNS/ATM Systems*, Appendix to the Report of the Fourth Meeting of the Special Committee for the Monitoring and Co-ordination of Development and Transition Planning for the Future Air Navigation System (FANS Phase II), Doc 9623, 1993.
28. ICAO, *Global Air Navigation Plan for CNS/ATM Systems*, Second Edition-2002, Doc 9750-AN/963.
29. The fourth and fifth editions seem to ignore the legal issues of GNSS, but they actually link themselves with the Appendix C of the third edition in 2007 in terms of legal considerations. *See* ICAO, *2013-2028 Global Air Navigation Plan*, Fourth Edition-2013, Doc 9750-AN/963, at 89; ICAO, *2016-2030 Global Air Navigation Plan*, Fifth Edition-2016, Doc 9750-AN/963, at 91.
30. ICAO, *supra* note 28, at 1-1-2.
31. Weber, *supra* note 5, at 98.
32. Diederiks-Verschoor & Kopal, *supra* note 3, at 69; Boakye Danquah Kofi Henaku, *The Law on Global Air Navigation by Satellite: A Legal Analysis of the ICAO CNS/ATM System* (AST Leiden, 1998), at xvii.

[C] How Is GNSS Technology Changing the Practice of International Air Law?

Only satellite technology could be capable of offering a homogenous service over large areas without the limitations of physical or geopolitical boundaries.[33] It is beyond the focus of this chapter to delve into too many deliberations on the technical and operational aspects of CNS/ATM systems. Suffice it to say that CNS/ATM systems consist mainly of the marriage between satellite technology and computers,[34] where GNSS constitutes a structural and essential component for both the introduction of PBN, and automatic dependent surveillance-broadcast (ADS-B).[35]

In bringing together the resources of outer space and airspace, the use of such spatial systems in aircraft operation and ANS provision inevitably generates provocative issues, under international air law, of legal significance.[36] Importantly, the aforementioned central role of GNSS makes it the primary focus of the legal implications of CNS/ATM systems.

On the one hand, different from the situation of traditional conventional air navigation facilities (*see* §4.04[B]), the substantial investment in constructing and operating GNSS leads most States to have to rely on GNSS controlled by others. In addition, concerns about continued availability based on a principle of non-discrimination are the major Achilles' heel of the implementation of CNS/ATM systems.[37]

On the other hand, GNSS is not always a reliable system, and its vulnerability in civil aviation applications is apparent (*see* §1.03).[38] This causes user States' to be concerned about the safety risk of their ANS based on GNSS.

Indeed, both of the two concerns above may connect with the issue of GNSS civil liability in the context of civil aviation, this due to State responsibility for the provision of air navigation facilities enshrined under Article 28 of the Chicago Convention. In practice, because of the inherent nexus with the safety of civil aviation, GNSS civil liability quickly became the significant element of the legal framework of CNS/ATM systems,[39] and it has thus been a primary focus of ICAO and its Member States as well (*see* §4.03).[40]

With lawmaking functions in the application as well as creation of international air law,[41] ICAO, more specifically its Legal Bureau, is obligated to provide appropriate

33. ICAO, *Resolution of the Legal and Institutional Aspects Associated with the Global Implementation of GNSS*, A35-WP/215, LE/17, 27/09/04, presented by the IATA.
34. Weber, *supra* note 5, at 97.
35. PBN is a navigation procedure, and ADS-B is a surveillance method under the concept of CNS/ATM systems; both of them are GNSS-based. *See* ICAO, *supra* note 11, at 1-3 to 1-4.
36. Abeyratne, *supra* note 4, at 35.
37. Ruwantissa Abeyratne, *Space Security Law* (Springer, 2011), at 23.
38. ICAO, *supra* note 11, at 5-1.
39. von der Dunk, *supra* note 10, at 130.
40. ICAO, *A Note on Legal Aspects of CNS/ATM, Including Views on How to Evaluate a Proposed Contractual Framework for GNSS*, A33-WP/188, LE/20, 26/9/01, presented by the United States, at 1.
41. Michael Milde, *International Air Law and ICAO* (Eleven International Publishing, 2012), at xiii & 3.

legal input on the above challenges and the new concerns of the international civil aviation community brought about by the use of GNSS. Further, ICAO recognises itself as "the only international organisation in a position effectively to coordinate global CNS/ATM activities."[42] In this light, ever since 1988, legal aspects of CNS/ATM systems (formerly known as the Future Air Navigation Systems) with the focus on GNSS have been in the Work Programme of the ICAO Legal Committee.[43] Due to the emergence of new issues of international aviation law such as the regulation of drones and the political fight between the user States – the Latin American States in particular – and provider States – which mainly refers to the US,[44] the priority of the said item fluctuated over the years from item 1 in 1992, to item 4 since 2015 up until today.[45] Although this current low priority may indicate that ICAO has washed its hands of the programme on legal aspects of CNS/ATM systems, the door to restart this matter nevertheless remains open.[46]

Back in the year 1992, ICAO already came to its preliminary conclusion that neither legal obstacles to the implementation of CNS/ATM systems, nor inconsistencies with the Chicago Convention exist.[47] Both GNSS provider States and user States have generally accepted this conclusion, and both of them have also agreed that no further work under international air law should delay the achievement of CNS/ATM systems.[48] The author does not intend to deny the above conclusion, but the *consistency* does not qualify as the *sufficiency* to deal with the new legal situation, particularly the issue of civil liability brought by the use of GNSS technology which differs significantly from the local traditional ground-based system (*see* §4.04[B]). The creation, in 1995, by the ICAO Council, of the LTEP is testimony to the author's viewpoint above. The LTEP is entrusted with finding a suitable solution for the legal

42. ICAO, *Appendix A General Policy*, to Consolidated statement of continuing ICAO policies and practices related to a global air traffic management (ATM) system and communications, navigation and surveillance/air traffic management (CNS/ATM) systems, ICAO Assembly Resolution A35-15, 2004.
43. Weber, *supra* note 5, at 99.
44. For example, the US called for the suspension on legal proposals on legal issues of CNS/ATM; however, the user States, with the support from the IATA, insisted that the efforts for a long-term legal framework must continue. *See* respectively: ICAO, *Assembly Resolution on a Practical Way Forward on Legal and Institutional Aspects of CNS/ATM*, A35-wp/216, LE/18, 28/9/04, presented by the United States of America [ICAO 216]; ICAO, *Considerations on the Final Report of the Secretariat Study Group on Legal Aspects of CNS/ATM*, A35-WP/179, LE/16, 21/9/04, presented by the 21 Member States of the Latin American Civil Aviation Commission [ICAO 179]; ICAO, *supra* note 33.
45. The changes in priority of legal aspects of CNS/ATM including GNSS since 1998 is as follows: 1988, Item 4; 1992, from Item 5 to Item 1; 2005, Item 3; 2013, Item 4; 2014, Item 5; 2015, Item 4. *See* Annual Reports of the ICAO Council of the above years.
46. UNIDROIT, *An Instrument on Third Party Liability for Global Navigation Satellite System (GNSS) Services: A Preliminary Study* (UNIDROIT, 2010), Study LXXIX – Preliminary Study, March 2010, at 51.
47. ICAO, *Report of the 28th Session of the ICAO Legal Committee*, Doc 9630-LC189, 1992.
48. ICAO, *World-Wide CNS/ATM Systems Implementation Conference Report*, May 1998, Doc 9719, at para. 5.3.1.

framework of GNSS with a two-stage approach, namely an appropriate approach for the immediate future, and a legal framework for the long-term future.[49]

After three meetings (in 1996, 1997, and 1998), the LTEP prepared the 'Draft Charter on the Rights and Obligations of States Relating to GNSS Service' (the 'Draft Charter') which was consequently adopted as Assembly Resolution A32-19, and other Recommendations for further study in 1998 (*see* §4.03[D]). The Draft Charter received wide support at the Worldwide CNS/ATM Systems Implementation Conference (Rio de Janeiro, 1998), and notably, the contribution of *regional* arrangements for the development of a global legal and institutional framework for the implementation of CNS/ATM systems including GNSS was also discussed at this conference.[50] Furthermore, Assembly Resolution A32-20 instructed the ICAO Council to establish a Secretariat Study Group on Legal Aspects of CNS/ATM Systems (the 'Group') at the end of 1998, so as to follow up the recommendations offered by the LTEP especially those concerning institutional issues and questions of civil liability, and to continue working on an appropriate long-term legal framework of GNSS including consideration of an international Convention for this purpose.[51] The Group submitted its progress report and final report to the thirty-third and thirty-fifth Session of the ICAO Assembly in 2001 and 2004 respectively.[52] Notably, the Group devoted substantial efforts in identifying possible solutions to the problem of civil liability, and the Group recognised several deficiencies and inadequacies of the ATC civil liability regime when dealing with GNSS civil liability (*see* §4.04[D]).[53] The Group also noted the unlawful interference with CNS/ATM systems, although not enough consideration on this topic was given.[54] Finally, the thirty-fifth Session of the ICAO Assembly approved the Group having accomplished its mission,[55] even though some States insisted that the Group should continue its task regarding legal aspects of CNS/ATM systems.[56] Thus, the mandate of ICAO on legal aspects of CNS/ATM systems including GNSS has only been referred to

49. The establishment of LTEP was decided by ICAO Assembly Resolution A31-7 after the recommendation by the 29th Session of ICAO Legal Committee. *See* ICAO, *Progress in the Work of the Panel of Legal and Technical Experts on the Establishment of a Legal Framework with Regard to GNSS (LTEP)*, A23-WP/24, LE/3, 18/6/98, at 1.
50. ICAO, *supra* note 48; *Ibid.*, at 3.
51. ICAO, *Development and Elaboration of an Appropriate Long-Term Legal Framework to Govern the Implementation of GNSS*, Assembly Resolution A32-20, 1998.
52. The thirty-fourth Session of ICAO Assembly was an extraordinary meeting mainly for the enlargement of the Council from 33 to 36 States in 2003. ICAO, *Provisional Agenda – Plenary Meetings*, A34-WP/1, P/1, 24/1/03.
53. ICAO, *Progress Report on the Establishment of a Legal Framework with Regard to GNS/ATM Including GNSS*, A33-WP/34, LE/5, 22/06/01, at 2 [ICAO 34]; ICAO, *Report on the Establishing of a Legal Framework with Regard to CNS/ATM Systems Including GNSS*, A35-WP/75, LE/5, 28/07/04, at A-5 [ICAO 75].
54. ICAO 34, *ibid.*, at 489; ICAO 75, *ibid.*, at A-52.
55. *See* ICAO, *A Practical Way Forward on Legal and Institutional Aspects of Communications, Navigation, Surveillance/Air Traffic Management (CNS/ATM) Systems*, A35-3, 8 October 2004.
56. ICAO, *Considerations on the Final Report of the Secretariat Study Group on Legal Aspects of CNS/ATM Systems: Need of a Binding Legal Framework*, A35-WP/179, LE/16, 21/9/04, at 4.

promote 'regional multinational organisms',[57] and monitor the development of contractual frameworks (*see* below).[58] Although at the thirty-sixth Session of the ICAO Assembly in 2007 there was a call for ICAO to set guidelines for this regional legal framework for the implementation of GNSS, nothing has been subsequently achieved in this regard.[59]

Among the above initiatives under the ICAO platform by the Council, Assembly, LTEP, or the Group, three approaches specific to GNSS civil liability have been proposed, as outlined below.[60]

(i) *Maintaining the current status.* The US ascertains that no indication has been found that the current liability regime under domestic law cannot cope with GNSS, and the efforts to establish a new universal liability system or a liability convention should be stopped as there is no consensus on any substantial change to the existing one.[61]
(ii) *Establishing a contractual framework.* The European States, mainly represented by Eurocontrol and the European Civil Aviation Conference (ECAC), favour the contractual framework where civil liability clauses are included[62] as an interim arrangement towards the new convention.[63]
(iii) *Concluding a new convention.* The African States and Latin-American States, as well as the European States, repeatedly confirm the need for a binding

57. The element of 'regional multinational organisms' was included in the consideration of a legal framework of CNS/ATM at the 36th Session of the ICAO Assembly in 2007. ICAO, *Regional Multinational Organisms*, A36-WP/231, LE/8, 18/9/07, presented by Colombia; ICAO, *Report of the Legal Commission for the General Section of Its Report and on Agenda Items 7, 8, 45, 46, 47 and 48*, A36-WP/341, P/49, 25/9/07, at 47-3.
58. This mandate of ICAO is described as follows:

> Invites Contracting States to also consider using regional organizations to develop mechanisms necessary to address any legal or institutional issues that could inhibit the implementation of CNS/ATM in the region, while ensuring that such mechanisms will be consistent with the Chicago Convention, and public international law; and
>
> Directs the Secretary General to monitor and, where appropriate, assist in the development of contractual frameworks to which parties may accede, inter alia, on the basis of the structure and model proposed by the Members of the European Civil Aviation Conference and the other regional civil aviation commissions, and on international law.

ICAO, *supra* note 55; ICAO, *Appendix F*, to Consolidated statement of continuing ICAO policies in the legal field, A36-26, 28 September 2007; ICAO, *Consolidated Statement of Continuing ICAO Policies in the Legal Field*, A37-22, 8 October 2010; ICAO, *Consolidated Statement of Continuing ICAO Policies in the Legal Field*, A39-11, 6 October 2016.
59. ICAO, *Evolution of Guidelines for Regional Legal Framework to Govern the Implementation of GNSS (SBAS)*, A36-WP/134, LE/6, 6/9/07, presented by India, at 3.
60. Although China and the Russia Federation are key GNSS providers, they seem to have kept silent on all proposals for GNSS civil liability.
61. ICAO 75, *supra* note 53, at 3. ICAO 216, *supra* note 44, at 4-5.
62. The clauses include but are not limited to the recognition of (strict) liability, compulsory risk coverage, recourse to arbitration, waiver of the right to invoke sovereignty immunity. ICAO, *Development of a Contractual Framework Leading Towards a Long-Term Legal Framework to Govern the Implementation of GNSS*, A35-WP/125, LE/11, 21/9/04, presented by the 41 Contracting States, Members of the European Civil Aviation Conference, at 3.
63. *Ibid.*, at 4.

convention on, among others, GNSS civil liability to address the clarity and legal certainty in the use of GNSS, and increase the confidence of user States.[64]

The controversial issue of GNSS civil liability has been the primary stumbling block to moving toward a long-term legal framework of CNS/ATM systems in the political game between provider States - which mainly refers to the US - and user States - including the European, Latin-American, and African States.[65] At this time, ICAO is moving very slowly by establishing a good 'excuse' in monitoring regional development. Evidently, most States, at least from the perspective of the *number* of States, have been calling for further action on GNSS civil liability, either by concluding a new convention for the long-term future, or adopting a contractual framework as an interim arrangement for the immediate future.[66] Based on the initiatives discussed above, the issue of GNSS civil liability is no longer novel for academia but it is still a complicated subject. Therefore, the author continues to examine whether present sources of international air law are adequate for the issue of GNSS civil liability; in other words, to see whether the viewpoint of the US is still tenable.

[D] Sources of International Air Law Concerning GNSS Civil Liability

Air law, aeronautical law, or aviation law, is defined as "a body of rules governing the use of airspace and its benefits for aviation, the traveling public, undertakings and the nations of the world."[67] Like space activities (*see* §3.02[B]), since its inception air transport is inherently international[68] as the high speed of an aircraft *en route* in the third dimension enables it to pass through the airspace of several States just to reach one single destination,[69] this being particularly true in the context of western Europe.[70] Air law is approaching its 100th anniversary, that is, if we take its birth to be the first public air law convention: the Paris Convention relating to the Regulation of Aerial Navigation of 1919.[71] Although modern air transport affects all sets of law, including both international law and municipal law, a large part of municipal air law is derived

64. *Ibid.*; ICAO, *Legal Aspects of GNSS*, AN-Conf/11-WP/143, 18/9/03, presented by the African States, at 2 & 3; ICAO 179, *supra* note 44, at 3.
65. Alessandra A.L. Andrade, *The Global Navigation Satellite System: Navigating into the New Millennium* (Ashgate, 2001), at 122.
66. Ludwig Weber & Jiefang Huang, *ICAO and GNSS*, 3(1) Outer Space Committee Newsletter 2000, at 43.
67. Pablo Mendes de Leon, *Introduction to Air Law* (Kluwer Law International, 2017), at 1.
68. Paul Stephen Dempsey, *Public International Air Law* (McGill University, 2008), at 1; Paul Stephen Dempsey, *Multilateral Conventions and Customary International Law*, in Paul Stephen Dempsey & Ram S. Jakhu (eds) Routledge Handbook of Public Aviation Law (Routledge, 2017), at 1.
69. Mendes de Leon, *supra* note 67, at 5.
70. Edward P. Warner, *International Air Transport*, 4(2) Foreign Affairs 1926, at 278.
71. *See* Pablo Mendes de Leon, *100 Years of Air Law – Turning the History of Air Law into Its Future: From Unilateral Acts to International Agreement Through Comity and Diplomacy*, in Stephen Hobe (eds), Air Law, Space Law, Cyber Law – the Institute of Air and Space Law at Age 90 (Carl Heymanns Verlag, 2016), at 25.

from international air law.[72] Considering the global nature of GNSS (*see* §1.02[E]) and the international characteristics of GNSS civil liability (*see* §2.05), this chapter mainly addresses international air law, but a summary of national air law on ATC civil liability is also discussed in §4.04[D].

Identifying the sources of international air law is not an easy task.[73] Simply speaking, however, these sources mainly refer to general principles of international law and international legal instruments, such as multilateral conventions and multi/bilateral agreements which govern the rights and duties of States with respect to control of their sovereign airspaces, and matters of aircraft safety, security, liability, and financing.[74] International air law may be further divided into two broad categories, namely Public and Private,[75] both of which are relevant to the issue of GNSS civil liability.

The fundamental principles of public international air law are mainly set out in one single primary source: the Chicago Convention (1944) which principally creates rights and obligations for States.[76] Strictly speaking, GNSS civil liability is a private law issue which shows no direct link with public international air law. This notwithstanding, as discussed above, the use of GNSS as a structural element of CNS/ATM systems falls within the scope of State responsibility concerning the provision of air navigation facilities regulated by Article 28 of the Chicago Convention (*see* §4.04[C]), while GNSS civil liability is the after-effect of the violation of that responsibility (*see* §2.03[A]).

Private international air law has been considered as one of the most advanced and prosperous instances of uniform private law[77] due to the achievement of unifying a series of rules for international carriage by air. This unification does not include every aspect of private international air law,[78] but mainly focuses on the regime of air carrier civil liability.[79] As GNSS civil liability is structured by either contractual liability (*see* §4.05[C]), or tort liability (*see* §4.05[D]), the treaties from Warsaw to Montreal via Rome are relevant sources. In addition, case law interpreting the application of these treaties as well as national aviation legislation is another critical source of private international air law.[80] No court cases concerning GNSS civil liability have appeared in

72. Chia-Jui Cheng, *New Sources of International Air Law*, in Chia-Jui Cheng (eds), The Use of Airspace and Outer Space for all Mankind in the 21st Century (Kluwer Law International, 1995), at 277.
73. *Ibid.*, at 279.
74. Bran F. Havel & Gabriel S. Sanchez, *The Principles and Practice of International Aviation Law* (Cambridge University Press, 2014), at 5.
75. Paul Stephen Dempsey, *The Future of International Air Law in the 21st Century*, 64 German Journal of Air and Space Law 2015, at 215; Ronald I.C. Bartsch, *International Air Law: A Practical Guide* (Ashgate, 2012), at 14.
76. Ludwig Weber, *Convention on International Civil Aviation - 60 Years*, 53(3) German Journal of Air and Space Law 2004, at 310; Milde, *supra* note 41, at 17.
77. Peter H. Sand, *The International Unification of Air Law*, 30(2) Law and Contemporary Problems 1965, at 400-401.
78. Paul Stephen Dempsey, *Aviation Liability Law* (LexisNexis, 2013), at 9.
79. George N. Tompkins, *Liability Rules Applicable to International Air Transportation as Developed by the Courts in the United States: From Warsaw 1929 to Montreal 1999* (Kluwer Law International, 2010), at 4.
80. Cheng, *supra* note 72, at 289.

practice so far,[81] yet case law on ATC civil liability particularly with a cross-border element, such as the 2002 Überlingen mid-air collision,[82] may be used as an analogy to see whether the use of GNSS could be viewed as a partial delegation of ATC service (*see* §4.04).

Furthermore, the distinction between hard law and soft law also applies to international air law,[83] even though sometimes the distinguishing line is not easy to identify in practice. This difficulty is particularly true for document systems made by public or private international aviation organisations such as ICAO and the IATA, as those internal documents are also recognised as one of the sources of international air law.[84] With regard to legal matters concerning GNSS, including the issue of civil liability, ICAO is rich in public statements and reports of the Council as well as its Legal Committee, Assembly Resolutions, SARPs, exchange of letters with States and other documents including manuals, GANPs and procedures, and most of them are sources of soft international air law (*see* §4.03). International air law today consists almost exclusively of written law.[85] However, in accordance with Article 38 of the Statute of the International Court of Justice, the sources of soft law are still probably to be identified as customary international air law with normative force, with the development of State practice particularly in the implementation of CNS/ATM systems.[86]

In short, sources of international air law concerning GNSS civil liability are abundant, including public and private, hard and soft, written, and also customary law. Nevertheless, whether those sources are adequate to deal with the issue of GNSS civil liability remains an open question so far which is further discussed in the following sections.

§4.03 STATE OF PLAY: THE ICAO REGIME FOR GNSS CIVIL LIABILITY

[A] The Mandate of ICAO on GNSS Civil Liability

Compared with international space law where no powerful institution mandated to regulate outer space activities exists,[87] one of the principal achievements made by international air law lies in the establishment of ICAO as the specialised agency of the

81. This does not mean no accident or economic loss has arisen by the use of GNSS if we expand our focus beyond aviation. *See* §1.03.
82. Überlingen mid-air collision happened under the control by Air Control Centre Zurich, Switzerland but above the territory of Germany. Therefore, there was a confusion on the liable party in this case. *See* German Federal Bureau of Aircraft Accidents Investigation, *Investigation Report*, Ax001-1-2/02, May 2004; Niels Arnoud van Antwerpen, *Cross-Border Provision of Air Navigation Services with Specific Reference to Europe: Safeguarding Transparent Lines of Responsibility and Liability* (Kluwer Law International, 2007), at 17.
83. Havel & Sanchez, *supra* note 74, at 12.
84. Mendes de Leon, *supra* note 67, at 4.
85. *Ibid.*, at 4-5.
86. Sofia Michaelides-Mateou, *Customary International Law in Aviation: A Hundred Years of Travel Through the Competing Norms of Sovereignty and Freedom of Overflight*, in Brian D. Lepard (eds), Reexamining Customary International Law (Cambridge University Press, 2017), at 313.
87. Atsuyo Ito, *Legal Aspects of Satellite Remote Sensing* (Martinus Nijhoff Publishers, 2011), at 20.

UN to offer global governance over international civil aviation.[88] Generally speaking, ICAO uses a two-pronged approach: a regulatory function and an implementing or enforcement function.[89]

Although it remains a tough question whether the regulatory function or legislative power of ICAO is "similar to a parliamentary lawmaking function, but limited to the sphere of civil aviation,"[90] ICAO has recognised its responsibility to consider the legal framework of GNSS with a focus on civil liability.[91] This responsibility is mainly undertaken by the Legal Committee, relevant bodies of the Assembly of ICAO, and its Secretariat - the Legal Affairs and External Relations Bureau[92] - which, among others, conducts research and studies, proposes new conventions and amendments of the Chicago Convention, and makes codifications in the field of private and public international air law.[93]

After decades of continuing efforts, ICAO has issued a series of regulatory and guidance materials related to GNSS, mainly in the form of Assembly Resolutions, SARPs, exchange of letters with States, and other documents including manuals, GANPs, and procedures. Whether the issue of GNSS civil liability in an air accident may be dealt by the role of the ICAO regime (not international air law as a whole) actually depends on the applicability and legal effect of those regulatory materials and guidance documents, either as direct legal sources, or indirect reference materials. The following sections analyse these topics further.

[B] Policy Statement and Guidance Documents

Policy statement refers to the declaration, commitment or expression of the management plans, intentions, and responsibility of ICAO.[94] On 9 March 1994, the ICAO Council released a policy document entitled 'Statement of Policy on CNS/ATM Systems implementation and operation' (hereafter refers to the 'Policy Statement') for the implementation of CNS/ATM systems including GNSS.[95] The Policy Statement outlines the most fundamental principles for the implementation of CNS/ATM systems, for example regarding the sovereignty and responsibilities of States, the continuity, and

88. Milde, *supra* note 41, at 128; Dempsey, *supra* note 68, at 43.
89. Jiefang Huang, *Aviation Safety through the role of ICAO: ICAO's Mechanisms and Practices* (Kluwer Law International, 2009), at 174.
90. *Ibid.*
91. Assad Kotaite, *ICAO's Role with respect to the Institutional Arrangements and Legal Framework of Global Navigation Satellite System (GNSS) Planning and Implementation*, XXI(II) Annals of Air and Space Law 1996, at 197.
92. The said Bureau reports to the Secretariat of ICAO, which provides technical, legal and administrative support to the Council and is headed by the Secretary General. *See* Mendes de Leon, *supra* note 67, at 36-37.
93. *See ibid.*, at 37; ICAO, *Legal Affairs and External Relations Bureau*, https://www.icao.int/secretariat/legal/Pages/default.aspx, last accessed 10 January 2018.
94. *Cf.* Section 1.1, Appendix 7. Framework for Safety Management Systems (SMS), Annex 6 of the Chicago Convention; Civil Aviation Department of the Government of the Hong Kong Special Administrative Region, *Safety Management Systems (SMS) for Air Operators and Maintenance Organizations: A Guide to Implementation (CAD 712)*, at 6.
95. *See* Appendix A to Chapter 2, in ICAO, *supra* note 28, at I-2-7 and I-2-8.

the quality of service of CNS/ATM systems. The role that this Policy Statement could play in the dispute settlement concerning GNSS civil liability is, however, unfortunately quite limited. First, the term 'policy statement' is "designed to inform rather than control,"[96] which means that a policy statement is not made legislatively; in other words, it is neither legislation, nor an interpretive rule.[97] The ICAO Council has the competence to take decisions, adopt policies, and release statements, but it does not hand down judicial decisions that are enforceable (Standards is an exception, *see* §4.03[E]).[98] Therefore, the Policy Statement is a non-binding norm and it lacks enforceability.[99] In this case, the Policy Statement could only create a moral, rather than mandatory obligation for *State* compliance.[100] Second, this moral obligation is actually directed only to the State operating CNS/ATM systems rather than GNSS provider States (e.g., the US, Russia, and China), and the regime of civil liability, which may be triggered in the breach of that moral obligation, is not GNSS civil liability but the well-known civil liability of ATC or ANS (*see* §4.04). Third, no specific rules concerning civil liability are directly found in the context of the Policy Statement.

Guidance materials are generally non-binding documents issued by ICAO to help illustrate or interpret the meaning of a requirement, specification or procedure regulated in technical codes and legal sources.[101] So far, guidance documents are usually connected with the issue of aviation safety, and they come in the form of Manuals, Circulars or GANPs.[102] With regard to GNSS, ICAO has issued several guidance materials, namely (i) GNSS Manual, which describes GNSS technology and operations so as to support State implementation decisions and planning;[103] and (ii) GANP, which serves as a worldwide reference to transform ANS in an evolutionary and inclusive manner.[104] Both the GNSS Manual and the GANP are updated periodically. Currently, the GNSS Manual is in its third version, released in 2017,[105] whereas the fifth edition of GANP was issued in 2016.[106] The GNSS Manual focuses only on a general description of GNSS, and it does not mention any content about the legal issues of GNSS, let alone the issue of civil liability. Although GNSS vulnerability is addressed in

96. Robert A. Anthony, *Interpretive Rules, Policy Statements, Guidances, Manuals, and the Like – Should Federal Agencies Use Them to Bind the Public?*, 41(6) Duke Law Journal 1992, at 1325.
97. *Cf. ibid.*
98. Ruwantissa Abeyratne, *Rulemaking in Air Transport: A Deconstructive Analysis* (Springer International Publishing Switzerland, 2016), at 86.
99. Kotaite, *supra* note 91, at 198 & 201; Alessandra Arrojado Lisboa de Andrade, *Navigating into the New Millennium: The Global Navigation Satellite System Regulatory Framework* (McGill LLM Thesis, 2000), at 84.
100. *Cf.* Ruwantissa Abeyratne, *Regulation of Air Transport: The Slumbering Sentinels* (Springer Science & Business Media, 2013), at 47.
101. Mendes de Leon, *supra* note 67, at 181 & 271.
102. *See* ICAO, *Outlook on ICAO Guidance Material*, https://www.icao.int/safety/Documents/Guidance%20Material.pdf, last accessed 11 January 2018.
103. ICAO, *supra* note 11, at 1-1.
104. ICAO, *ICAO's Global Air Navigation Plan*, https://www.icao.int/airnavigation/Documents/GANP_at_glance_flyer.pdf, last accessed 12 January 2017.
105. ICAO, *Products & Services Catalogue 2017*, https://www.icao.int/publications/catalogue/cat_2017_en.pdf, last accessed 14 January 2018.
106. ICAO, *Doc Series: Doc 9750 Global Air Navigation Plan*, https://www.icao.int/publications/Pages/Publication.aspx?docnum=9750, last accessed 12 January 2017.

the GNSS Manual, this fact could only prove that aviation users of GNSS are aware of the risk of application, and it does not raise any legal significance on the issue of GNSS civil liability. Different to the GNSS Manual, the second edition (2002) of the GANP, entitled 'Global Air Navigation Plan for CNS/ATM Systems', provides practical guidance on the technical, operational, economic, financial, legal, and institutional elements of CNS/ATM implementation.[107] Chapter 11 of the GANP deals with 'legal issues', but the materials codified in this chapter are "intended to assist States in identifying relevant legal issues, which they may encounter in the planning and implementation of CNS/ATM systems."[108] In this sense, Chapter 11 of the GANP including its annexes is merely a collection of ICAO policies and documents concerning legal considerations of CNS/ATM systems addressing issues mainly related to GNSS.[109] The predominant viewpoint holds that the GANP is merely a high-level 'policy' or a 'strategic' document endorsing the global strategic directions for safety and air navigation.[110] Therefore, the GANP has neither the intention, nor the ability to work as a code of legal rules concerning the implementation of GNSS.

At this juncture it could be briefly concluded that neither policy statements, nor guidance materials issued by ICAO have any legal effect, and they do not qualify as legal sources but only as reference materials which could be used in case of an air accident caused by the malfunction of GNSS. The author holds that all the responsibilities and obligations related to GNSS implementation in those materials are actually requested by ICAO or its Council to the Member States from the perspective of *public* management, rather than the GNSS provider States' commitments to the users or third parties which may serve as a legal basis for a claim for compensation. Therefore, it is inappropriate for victims to use those reference materials as legal sources to claim compensation under *private* law.

[C] Exchange of Letters

Recommended by the tenth Air Navigation Conference held in 1988,[111] the ICAO Council requested the Secretary General to conclude an agreement with GNSS provider States.[112] The Legal Committee of ICAO then prepared a draft agreement where the provision of 'Responsibility and *Liability* for Service' was incorporated as follows:

107. Weber, *supra* note 5, at 98.
108. ICAO, *supra* note 28, at I-11-1.
109. *Ibid.*
110. Mendes de Leon, *supra* note 67, at 27; Francis Schubert, *Air Navigation*, in Paul Stephen Dempsey & Ram S. Jakhu (eds), Routledge Handbook of Public Aviation Law (Routledge, 2017), at 101; Victor Iatsouk, *Development of Standards for Aeronautical Satellite Navigation System*, 54 Acta Astronautica 2004, at 961.
111. ICAO, *Report of the Tenth Air Navigation Conference*, Doc. 9583, AN-CONF/10, at 4.7; Andrade, *supra* note 99, at 84-85.
112. *See* ICAO, *Draft Agreement Between the International Civil Aviation Organization (ICAO) and GNSS Signal Provider Regarding the Provision of Signals for GNSS Services*, in Doc. 9630-LC/189, at para. 3:38.10; Andrade, *supra* note 99, at 84-85.

> [Name of the provider of the GNSS signal] shall be responsible and *liable* to take all necessary measures to maintain the integrity and reliability of the GNSS signal and its continuous and uninterrupted availability in order to meet the needs of air navigation [...] .[113]

However, in order to avoid complex domestic political consequences,[114] the US government sent ICAO a letter, rather than an agreement, in 1994 (updated in 2007).[115] The Russian government exchanged similar letters with ICAO in 1996.[116] ICAO accepted and communicated the exchanges of letters to its Member States,[117] and recognised them as a transitional arrangement compared with the conclusion of an agreement or convention, which was considered as the final form for GNSS legal matters[118] with GNSS provider States.[119]

The letters exchanged between ICAO and provider States reiterated certain fundamental principles described formerly in the Policy Statement (*see* §4.03[B]) and the draft agreement above-mentioned,[120] which are relevant to the continuous and non-discriminatory worldwide access, free of direct user fees, and compliance with the performance parameters (*see* §1.02[D]) required by ICAO SARPs (*see* §4.03[E]).[121] Differently, the provision on the liability of service was removed, or perhaps ignored on purpose,[122] because GNSS provider States, that are the US and the Russian Federation, viewed the burden of undertaking liability as too onerous, and the issue of liability as too complex to be dealt within the draft agreement and subsequent letters.[123] Nevertheless, those letters indeed qualify as a quality and continuity commitment for the provision of GNSS signals made by the US and Russian governments to ICAO, or the international civil aviation community as a whole.[124] Two questions arise here, one

113. ICAO, *ibid.*
114. Under the US legal system, to execute a formal executive agreement may have to follow the formal procedure of concluding a treaty; international law does not differentiate between treaties and executive agreements. *See* Congressional Research Service Library of Congress, *Treaties and Other International Agreements: The Role of the United States Senate* (U.S. Government Printing Office Washington, 2001), at 4; Jonathan M. Epstein, *Global Positioning System (GPS): Defining the Legal Issues of Its Issues of Its Expanding Civil Use,* 61 Journal of Air Law and Commerce 1995, at 275.
115. *See* ICAO 75, *supra* note 53, at Attachment B; FAA, *2007 GPS and WAAS Service Commitments to ICAO,* https://www.gps.gov/policy/cooperation/icao/2007-service-commitments.pdf, last accessed 14 January 2018.
116. *See* ICAO 75, *supra* note 53, at Attachment C.
117. Henaku, *supra* note 32, at 180.
118. ICAO, *Report of the Rapporteur on the 'Consideration, with Regard to Global Navigation Satellite System (GNSS), of the Establishment of a Legal Framework'*, LC/29-WP/3-1, 3 March 1994, presented by Kenneth Rattray, at Annex III.
119. Weber, *supra* note 5, at 99.
120. ICAO 75, *supra* note 53, at A-4.
121. *See* ICAO 75, *supra* note 53, at Attachment B & C.
122. Bergamasco, *infra note* 196, at 2.
123. *See* ICAO, *Report of the 29th Session of the ICAO Legal Committee,* in Doc. 9630-LC/189, at para. 3:38.7.2.
124. Kotaite, *supra* note 91, at 202.

    A report issued by RAND stated that the letters sent by the US government to ICAO was "notably not a commitment", and addressed that "the US has not entered into any commitment to provide GPS services to particular parties or to agreed-upon specifications." This argument is obviously untenable because the updated letter in 2007 states that: "the US Government

relating to what the legal consequences would be if the US or the Russian government breached such commitment, and the other relating to whether victims are entitled to claim compensation if damage were caused by such a breach. The answers depend on the legal significance of those letters, which in turn has further led to a variety of questions from different perspectives.[125]

First, does ICAO have the competence to make agreements with others regarding the matter of GNSS? One opinion holds that international organisations are entitled to conclude international agreements, and some practice and judicial decisions have endorsed this opinion.[126] Therefore ICAO, as an international organisation regulating international civil aviation, has both the general and specific capacity to enter into international agreements governing the use of GNSS for air navigation.[127] An opposite viewpoint insists that ICAO lacks legal powers and constitutional standing to make a legally binding understanding or a formal agreement concerning GNSS in the name of the international civil aviation community.[128] Half of the second viewpoint receives this author's support. It has to be admitted that ICAO is entitled to enter into agreements with other international bodies according to Article 65 of the Chicago Convention;[129] however, this provision is only for the purpose of facilitating the administrative work of ICAO, and it would be impermissible to extend its applicability to the provision of GNSS signals.[130] Besides, ICAO is not a supranational organisation like the EU, and its Member States do not transfer their sovereign right of concluding agreements or treaties to ICAO. To say the least, even though the competence of ICAO to make agreements on the use of GNSS is recognised by certain scholars, the question on whether those letters have legal effect depends on the contents of those letters and the intentions of the US and Russian governments which may be requested to undertake the obligation of compensation for damage caused by GNSS, rather than that of ICAO (*see* below).

---

maintains its commitment to [...] ", which means that the letters in both 1996 and 2007 qualify the form of a commitment. *See* respectively: Scott Pace & Gerald Frost, et al., *The Global Positioning System: Assessing National Policies* (Rand, 1995), at 42 & 43; FAA, *supra* note 115.

125. *See* Andrade, *supra* note 99, at 88-91; Henaku, *supra* note 32, at 182-183; Epstein, *supra* note 114, at 274-277.

126. *See* Henry G. Schermers & Niels Blokker, *International Institutional Law: Unity Within Diversity* (M. Nijhoff, 1995), at 1096; International Court of Justice, *Reparation for injuries suffered in the service of the Nations*, Advisory Opinion, [1949] ICJ Rep 174.

127. Henaku, *supra* note 32, at 182.

128. ICAO, *supra* note 123, at para. 3:31; Michael Milde, *Solutions in Search of a Problem? Legal Problems of the GNSS*, XXII(2) Annuals of Air and Space Law 1997, at 201, 203 & 209; Michael Milde, *Institutional and Legal Problems of the Global Navigation Satellite System (GNSS) – Solutions in Search of a Problem*, in Chia-Jui Cheng & Doo Hwan Kim (eds), The Utilization of the World's Air Space and Free Outer Space in the 21st Century (Kluwer Law International, 2000), at 352.

129. Article 65 of the Chicago Convention reads as follows:

> The Council, on behalf of the Organization, may enter into agreements with other international bodies for the maintenance of common services and for common arrangements concerning personnel and, with the approval of the Assembly, may enter into such other arrangements as may facilitate the work of the Organization.

130. Milde, *supra* note 128, at 201.

Second, are the exchanges of letters unilateral or one part of bilateral acts? Although this question does not directly answer what the legal significance of those letters is, it deserves particular attention as the majority of viewpoints seem to indicate that an exchange of letters constitutes a *unilateral* promise. The author does not share this viewpoint. The official statement made by former President Reagan after the downing of Korean Air Flight 007 due to navigation errors, constitutes a purely unilateral commitment without any party's request.[131] Differently, however, the communication concerning the use of GNSS was *interactive* and the letters were *exchanged* between ICAO and the US and Russian governments. In private law, a contract is made by an offer from Party A and an acceptance by Party B;[132] similarly, the US and Russian governments made the offer on GNSS service commitment, and ICAO, through its Council, accepted that offer, or vice versa. Therefore, the author argues that the exchanges of letters actually constitute one form of Memorandum of Understanding (MOU), or a *political* (versus *legal*) agreement, as named by someone,[133] which is non-binding and not enforceable in courts.[134]

Third, do those letters have any binding force of law on the US and Russian governments? The author agrees that the title of an instrument is not decisive for its legal significance, and what are of paramount importance to answer the above question are the contents of those letters.[135] Throughout ICAO's history, it is not uncommon to use letters for the purpose of communication between ICAO and Member States on some issues such as the status of war or national emergency,[136] but few of them qualify as legal agreements or international treaties.

Although the exchanges of letters or any other means could satisfy the *form* of an agreement or a treaty,[137] two crucial factors have to be examined: (i) the intention to conclude a treaty with reference to the *substance*, not the *form*, of the agreement; and (ii) the clear and specific description of the legal obligations of the parties.[138]

On the one hand, the letters exchanged between ICAO and the US and Russian governments do not satisfy the first criteria.[139] For example, the letters sent by the US government in 1994 and 2007 were submitted 'in lieu of an agreement', and comprise 'mutual understandings', and 'political commitments';[140] the letter sent by the Russian government in 1996 clarified that "we are prepared to conclude an agreement with

131. Pace, Frost, et al., *supra* note 124, at 42.
132. Ewan McKendrick, *Contract law: Text, Cases, and Materials* (Oxford University Press, 2014), at 43.
133. *See* Congress, *supra* note 114, at 4; Epstein, *supra* note 114, at 276.
134. *Ibid.*
135. ICAO, *supra* note 123, at para. 3:28.
136. *See* Jiefang Huang, *Aviation Safety and ICAO* (Leiden PhD Thesis, 2009), at 94-95.
137. Mendes de Leon, *supra* note 67, at 6.
138. Congress, *supra* note 114, at 3-4.
139. Another view is that the Russian-ICAO exchange of letters is different with the US-ICAO one, as the former contains "a clear expression of an intention to be bound by the terms agreement." The author does not share this viewpoint (*see* the discussion below). *See* Henaku, *supra* note 32, at 183.
140. *See* ICAO 75, *supra* note 53, at Attachment B; FAA, *supra* note 115.

ICAO for the use of the GLONASS system [...]," which in turn means that this letter was not yet an agreement.[141]

On the other hand, it is arguable whether those letters describe the detailed rules on the obligation to provide GNSS signals,[142] even though it could be confirmed easily that no specific obligation concerning compensation is mentioned. Therefore, the exchanges of letters between ICAO and the US and Russian governments do not represent an agreement with *legally* binding effect in the normal sense of international law.[143]

Nevertheless, the lack of a legally binding effect does not decrease the moral or political obligations of the US and Russian governments to provide GNSS signals in a good way which complies with their quality commitment stated in those letters, bearing in mind that an act of a State is fairly serious and a public commitment of a State should be very genuine. Nevertheless, those letters cannot directly serve dispute settlement concerning GNSS civil liability.

[D] Assembly Resolutions

ICAO Assembly Resolutions were often used to include other forms of regulatory materials such as draft treaties, model legislation, and model clauses.[144] So far ICAO Assembly Resolutions have almost addressed all aspects of international civil aviation, including the legal framework of GNSS.[145] After a series of discussions of the LTEP in the late 1990s, the thirty-second session of the ICAO Assembly adopted two resolutions on GNSS, namely: (i) Charter on the rights and obligations of States relating to the GNSS services (A32-19, the 'Charter'); (ii) Development and elaboration of an appropriate long-term legal framework to govern the implementation of GNSS (A32-20).[146] As one of the first tangible results of ICAO's efforts,[147] Resolution A32-19 'solemnly declares' several fundamental principles on the implementation and operation of GNSS, most of which are, or intend to be imposed on States, including both GNSS provider States and user States.[148] Further, Resolution A32-19 was followed by some Recommendations offered by the LTEP on those subjects which needed to be further studied before a consensus was reached;[149] notably, those recommendations were not included in Resolution A32-19.[150] Resolution A32-20 confirmed ICAO's recognition of the importance of GNSS legal issues, and 'instructs' the Council and the Secretary

141. *See* ICAO 75, *supra* note 53, at Attachment C.
142. Epstein, *supra* note 114, at 275.
143. *See* Andrade, *supra* note 99, at 91; Henaku, *supra* note 32, at 183; Epstein, *supra* note 114, at 276.
144. Huang, *supra* note 136, at 182.
145. Mendes de Leon, *supra* note 67, at 35.
146. ICAO, *Assembly Resolutions in Force (as of 6 October 2016)*, Doc 10075, at V-9 to V-11.
147. Mendes de Leon, *supra* note 67, at 306.
148. ICAO, *supra* note 146, at V-9 to V-11.
149. Henaku, *supra* note 32, at 245.
150. ICAO, *supra* note 112, at 3.

General to take further actions on the law of GNSS, particularly the recommendations offered by the LTEP.[151]

ICAO Assembly Resolutions may be categorised as either 'internal' or 'external' rules according to their content: the former refers to resolutions concerning ICAO's structure, functioning or procedures; the latter to resolutions directly addressing Member States.[152] The legal effect of resolutions differs between internal and external resolutions. Even though the general opinion is that internal resolutions made by ICAO – as a special agency of the UN – may be binding, they are only for the operation of ICAO. Resolution A32-20 constitutes an internal resolution because it merely regulated the Assembly's instruction or call for action to the Council and the Secretary General, rather than the legal obligation of GNSS provider States, and this reduces its status as a legal source of GNSS civil liability in court. Therefore, the author devotes his attention more to the legal significance of Resolution A32-19 – which qualifies as an external resolution – and its follow-up recommendations, and their potential role when dealing with the issue of GNSS civil liability.

Although Resolution A32-19 does not fully answer the matter of civil liability in a direct way,[153] its fundamental principles are more or less of a relationship with the regime of GNSS civil liability. For example, one of the fundamental principles reads as follows:

> Every State providing GNSS services, including signals, or under whose jurisdiction such services are provided, shall ensure the continuity, availability, integrity, accuracy and reliability of such services, including effective arrangements to minimize the operational impact of system malfunctions or failure, and to achieve expeditious service recovery. Such State shall ensure that the services are in accordance with ICAO Standards. States shall provide in due time aeronautical information on any modification of the GNSS services that may affect the provision of the services.[154]

This provision directly addresses the obligations of GNSS provider States, including the obligation that those States *shall* comply with the relevant quality requirements of GNSS signals. If an air accident happened because the provider State did not fulfil those requirements, could a claim for compensation against that State be raised? The answer depends on whether Resolution A32-19 is legally binding on GNSS provider States.

The legal effect of Assembly Resolutions made by supreme bodies of international organisations has to be determined according to the relevant principles of general international law, and ICAO shares the same situation.[155] After a long and fierce dispute, the emerging consensus on the above issue prefers to distinguish between different resolutions according to a number of factors, such as the intention of

151. ICAO, *supra* note 146, at V-10 to V-11; Weber, *supra* note 5, at 101.
152. *Cf.* Huang, *supra* note 136, at 182.
153. Lesley Jane Smith, *Legal Aspects of Satellite Navigation*, in Frans von der Dunk with Fabio Tronchetti, Handbook of Space Law (Edward Elgar Publishing, 2015), at 604.
154. ICAO, *supra* note 146, at V-10.
155. Mendes de Leon, *supra* note 67, at 35.

the Assembly, the content of principles, and the majority in favour of adoption.[156] Accordingly, Professor Jiefang Huang categorised ICAO Assembly Resolutions as follows:

- *Declaratory resolutions*, which declare or confirm customary international rules as evidenced by the votes and views of States.
- *Interpretative resolutions*, which provide interpretation to existing treaty provision or application.
- *Pre-legislation resolutions*, which ICAO adopts as a means to authenticate a multilateral treaty, or prepare model legislation or model clauses for a domestic law or a bilateral treaty.
- *Directive resolutions*, commonly called 'internal rules' (*see* above), which give instructions to subordinate bodies.[157]

Resolution A32-19 seems to qualify as a pre-legislation resolution. First, the use of GNSS in civil aviation was a new concept back then when no customary international rules concerning GNSS had been developed. Hence, the fundamental principles regulated in Resolution A32-19 constitute neither a declaratory, nor an interpretive resolution.[158] Besides, since the subjects which Resolution A32-19 addresses are the Member States rather than ICAO's sub-bodies, it does not comply with a directive resolution. Back in the late 1990s, the LTEP identified many legal forms to regulate those fundamental principles for the use of GNSS, including a convention, a charter, multilateral or bilateral agreements, SARPs, and an Assembly or a Council resolution/declaration.[159] After a protracted discussion, the LTEP decided that a charter should be drafted with the effect of either an international convention, or an Assembly Resolution.[160] Although the majority of the LTEP were in favour of an international convention, the draft charter was finally submitted to the thirty-second Assembly for adoption as an interim arrangement for the conclusion of a convention.[161] The above fact, in turn, demonstrates that Resolution A32-19 is purely an Assembly Resolution without the legal effect of an international convention, though it is a preparation for a future international convention.

Actually, the methods to judge the legal significance of exchanges of letters (*see* above) are also relevant for determining the force of ICAO Assembly Resolutions.[162] Resolution A32-19 regulates certain policy statements rather than concrete rules of law, so it does not bind ICAO Member States like a treaty.[163] Those principles do not differ to any great extent from the Policy Statement and exchanges of letters as discussed

---

156. Hermann Mosler, *The International Society as a Legal Community* (Brill, 1980), at 88-89.
157. Huang, *supra* note 136, at 182.
158. At present, the most recognised customary international air law refers to the principle of complete and exclusive sovereignty of States over the airspace above their territory, and this is respected in ICAO Assembly Resolution A32-19.
159. Henaku, *supra* note 32, at 242.
160. *Ibid.*
161. ICAO, *supra* note 49, at 2.
162. Mendes de Leon, *supra* note 67, at 6.
163. Paul B. Larsen, Joseph Sweeney & John Gillick, *Aviation Law: Cases, Laws and Related Sources* (Martinus Nijhoff Publishers, 2012), at 1050.

above.[164] Even though the language used in Resolution A32-19 seems somewhat mandatory,[165] and the title of 'Charter' shows great importance and an indication of it being a legal instrument,[166] the Member States of ICAO never expressed an intention to accept this resolution as an international convention. Therefore, the predominant view is that the Charter is not legally binding.[167] However, the non-binding effect of Resolution A32-19 does not indicate that it is useless. Resolution A32-19 was adopted by a majority vote, and this creates presumptions of binding commitments to some extent, at least for those States which voted in favour.[168] Again, these kind of 'commitments' are usually political or moral, as "the attitude of States towards a given resolution (or a particular rule set forth in a resolution), expressed by vote or otherwise, is often motivated by political or other non-legal considerations."[169]

In the author's opinion, the vote for the adoption of Resolution A32-19 is like the 'signature' in the treaty-making process, and it merely showed the willingness of the signatory State to proceed for the ratification, acceptance or approval in good faith.[170] However, before the ratification, acceptance or approval is finished according to a domestic legitimate procedure, the signature or the adoption of a resolution does not confer any legal effect, and this is also why the author labelled Resolution A32-19 as 'pre-legislation resolution' (*see* above).

At this point, one question that may arise is whether Resolution A32-19 is able to acquire the status of customary international law. Although Resolution A32-19 did not qualify as a declaratory resolution which declares or restates customary international law back to the time of its adoption, it could be viewed as a starting point of customary international law on the provision of GNSS. Since its adoption, Resolution A32-19 has been playing a pivotal role in the formation of customary international law.[171] Resolution A32-19 itself cannot create a rule of customary law, but the fundamental principles regulated by it may be developed as sources of customary international law with the increasing use of GNSS in civil aviation practice, through consistent and repeated de facto acceptance by States as determining their freedom of action.[172] Generally speaking, to identify a rule of customary international law, two criteria are

164. Henaku, *supra* note 32, at 244.
165. Nie Jingjing, *The Future of Uniform International Rules on GNSS Liability*, 54 Proceedings of the International Institute of Space Law 2011, at 343; Andrade, *supra* note 65, at 89;
166. Milde, *supra* note 128, at 209.
167. *Ibid.*, at 210; Andrade, *supra* note 99, at 94; Jingjing, *supra* note 165, at 343; von der Dunk, *supra* note 23, at 28.
168. *Cf.* Huang, *supra* note 136, at 185; United Nations International Law Commission, *Yearbook of the International Law Commission 1998: Documents of the Fiftieth Session (Volume 2 Part 1)* (United Nations Publications, 2009), at 336.
169. *Military and Paramilitary Activities in and Against Nicaragua* (*Nicaragua v. United States of America*), Merits, Judgment, I.C.J. Reports 1986, at 99; UN, *Report of the International Law Commission: Sixty-Eighth Session*, 2 May-10 June and 4 July-12 August 2016, A/71/10, at 108.
170. Articles 10 and 18 of the Vienna Convention on the Law of Treaties (1969); United Nations Library, *What Is the Difference Between Signing, Ratification and Sccession of UN Treaties*, http://ask.un.org/faq/14594, last accessed 23 January 2018.
171. Michael Rycroft (Eds), *Satellite Navigation Systems: Policy, Commercial and Technical Interaction* (Springer Science + Business Media Dordrecht, 2003), at 196.
172. von der Dunk, *supra* note 23, at 28.

essential as concluded by the ILC: (i) a general State practice; and (ii) acceptance as law (*opinio juris*).[173]

Throughout decades of use of CNS/ATM systems, the author believes that the fundamental principles regulated in Resolution A32-19 have been followed not only by GNSS provider States, but also by user States under whose jurisdiction such services are provided. However, whether those States view these fundamental principles as *legal* obligations is questionable. For example, in the updated letters exchanged with ICAO in 2007, the US government clarified that its commitment to ICAO on the provision of GNSS service or signals is merely a 'political' commitment,[174] which indicates that the US government did not accept it as a 'legal' obligation. In June 2016, China released a policy document titled 'China's BeiDou Navigation Satellite System', which declared that "China is committed to [...] providing continuous, stable and reliable open services to users free of charge."[175] This publication is however purely a political 'white paper' which does not show any 'legal opinion' on the use of China's GNSS, let alone the issue of civil liability. Therefore, at the current stage the status of customary international law of those fundamental principles regulated by Resolution A23-19 remains open, and it is far from being 'a general practice that is accepted as law (*opinio juris*)' worldwide.[176]

As for the recommendations following Resolution A32-19, their value to the dispute settlement concerning GNSS civil liability remains negative. The recommendations were not included in Resolution A32-19.[177] Recommendations 9-11 address the issue of civil liability, which outline key terms to be further defined, rules of evidence to be regulated, and several questions to be answered.[178] Those recommendations are not legal rules of GNSS civil liability but rather *topics* requiring further study for an appropriate legal regime of GNSS civil liability which could not reach consensus in the form of Resolution A32-19.[179] Therefore, the follow-up recommendations do not work as a legal basis or source of GNSS civil liability in the process of dispute settlement.

[E] Standards and Recommended Practices and Procedures for Air Navigation Services

To promote the safe and orderly development of civil aviation, the ICAO Council is mandated to adopt and update SARPs, which are viewed as a flexible means to secure the highest practicable degree of uniformity in aspects such as regulations, standards, and procedures related to air navigation systems, including those based on GNSS.[180] Furthermore, the ICAO Council *approves* Procedures for Air Navigation Services

173. UN, *supra* note 169, at 76.
174. FAA, *supra* note 115.
175. The State Council Information Office of the People's Republic of China, *China's BeiDou Navigation Satellite System* (Foreign Language Press, 2016), at 9.
176. UN, *supra* note 169, at 76.
177. Henaku, *supra* note 32, at 245.
178. *See* ICAO, *supra* note 49, at B-3 to B-4.
179. Kim Murray, *The Law Related to Satellite Navigation and Air Traffic Management Systems – a View from the South Pacific*, 31 VUW Law Review 2013, at 399.
180. Article 37 of the Chicago Convention.

(PANS), comprising materials considered immature or too detailed for SARPs, as well as operating practices for worldwide application.[181] SARPs are grouped into Annexes to the Chicago Convention 'for convenience'.[182] The terms 'Standards' and 'Recommended Practices' are not defined in the Chicago Convention, but their definition may be found in several resolutions and Annexes;[183] PANS are classified as 'guidance material',[184] yet this term is not even expressly provided for in the Chicago Convention.[185] The main difference among the terms Standards, Recommended Practices and PANS lies in the degree of necessity for unification, namely: (i) in order to qualify as standards, the unification application must be considered *necessary*;[186] (ii) in order to qualify as Recommended Practice, the unification application must be considered *desirable, but not essential*;[187] and (iii) in order to qualify for PANS status, the procedure must be agreed as *suitable* for application on a worldwide basis.[188] As of January 2018, there are nineteen annexes, and seventeen of them lay down rules of conduct of a technical nature.[189]

The implementation of GNSS gave rise to the need to amend certain ICAO SARPs,[190] therefore in 1993 the ICAO Air Navigation Commission established the Global Navigation Satellite System Panel (GNSSP), subsequently renamed Navigation System Panel (NSP), with that basic objective.[191] In 2001, the first package of SARPs, supporting GNSS operations based on augmenting core satellite constellation signals to meet safety and reliability requirements,[192] was introduced in Volume I (Radio Navigation Aids) of Annex 10 (Aeronautical Telecommunications) to the Chicago

---

181. Milde, *supra* note 41, at 174; ICAO, *Directives to Divisional-Type Air Navigation Meetings and Rules of Procedure for Their Conduct*, Doc 8143-AN/873/3, 1983, at 10; Mendes de Leon, *supra* note 67, at 294.
182. Article 54 (i) of the Chicago Convention; Milde, *supra* note 41, at 167.
183. The terms Standard and Recommended Practices are defined as follows:

> *Standard:* Any specification for physical characteristics, configuration, material, performance, personnel or procedure, the uniform application of which is recognised as necessary for the safety or regularity of international air navigation and to which Contracting States will conform in accordance with the Convention; in the event of impossibility of compliance, notification to the Council is compulsory under Article 38.
>
> *Recommended Practice:* Any specification for physical characteristics, configuration, material, performance, personnel or procedure, the uniform application of which is recognised as desirable in the interest of safety, regularity or efficiency of international air navigation, and to which Contracting States will endeavor to conform in accordance with the Convention.

ICAO, *Definition of International Standards and Recommended Practices*, A1-31, 1947; ICAO, *Volume I: Radio Navigation Aids*, Annex 10 to the Chicago Convention, at viii.
184. *See* ICAO, *supra* note 102.
185. Milde, *supra* note 41, at 175.
186. ICAO, *supra* note 181, at 6.
187. *Ibid.*, at 7.
188. *Ibid.*, at 10.
189. ICAO, *Making an ICAO Standard*, https://www.icao.int/safety/airnavigation/Pages/standard.aspx, last accessed 30 January 2018; Abeyratne, *supra* note 98, at 83.
190. Weber, *supra* note 5, at 99.
191. Iatsouk, *supra* note 110, at 962.
192. ICAO, *supra* note 11, at vi.

Convention which became applicable as of November 2001.[193] Since then, ICAO has been discharging the responsibility of GNSS SARPs updates and enhancements in accordance with Article 37 of the Chicago Convention.[194] In addition to Annex 10, many other annexes and technical documents have been under way to accommodate proper introduction of GNSS into civil aviation,[195] including but not limited to: Annex 2 (Rules of the Air), Annex 4 (Aeronautical Charts), Annex 6 (Operation of Aircraft), Annex 11 (Air Traffic Services), Procedure design standards in the Procedures for Air Navigation Services - Aircraft Operations (PANS-OPS), and Procedures for Air Navigation Services - Air Traffic Management (PANS-ATM).[196] The above SARPs and PANS provide performance assurances, though to a varying degree, for the implementation of GNSS for the purpose of air navigation.[197]

However, the majority of the above instruments constitute a comprehensive *technical* safety code for civil aviation,[198] and their objective is to achieve compatibility and uniformity between States on the implementation of GNSS in their jurisdictions from a more *technical* point of view, rather than to regulate *legal* obligations and rights of GNSS providers or users. Therefore, it is argued that SARPs and PANS are not the appropriate instruments to deal with the legal issue of GNSS civil liability as they are traditionally used to regulate *technical* matters.[199] The author gives his support to the above argument. The technical regulations and guidance materials may nonetheless play a *supporting* role of importance when dealing with the issue of GNSS civil liability, through either Article 28 or Article 38 of the Chicago Convention.

On the one hand, Article 28 of the Chicago Convention requests each contracting State to provide air navigation facilities in its territory "in accordance with the standards and practices recommended or established from time to time," and this provision's function is a general responsibility on ANS required from all Member States of ICAO, regardless of whether they are GNSS provider States, or user States. GNSS is a critical element of radio navigation aids in the provision of ANS under CNS/ATM system (*see* §4.02[B]), and the States providing ANS have the responsibility to not only oversee the performance of GNSS used in their territory by means of legislation or other governance tools, but also to ensure that GNSS complies with the parameters (*see* §1.02[D]) regulated as *Standards* if someone argues the weak binding force of Recommended Practices and PANS (*see* below). In cases where damage was caused by the failure of the provision of ANS due to the GNSS non-compliance with ICAO Standards in their territory, the civil liability regime of ATC or ANS will apply, which is different from the regime of GNSS civil liability (*see* §4.04).

193. ICAO, *supra* note 11, at ix; Kotaite, *supra* note 91, at 199.
194. Mohamed Smaoui, *ICAO Global Provisions and Regional Developments Related to GNSS, Presented to Joint ACAC/ICAO MID Workshop on GNSS* (Rabat, Morocco, 7-8 November 2017).
195. von der Dunk, *supra* note 23, at 23.
196. Federico Bergamasco, *GNSS Liability: Current Legal Framework and Perspectives for the Future from the International Aviation Point of View*, 4 The Aviation & Space Journal 2014, at 3.
197. Weber, *supra* note 5, at 102.
198. Huang, *supra* note 136, at 44.
199. *Cf.* Kotaite, *supra* note 91, at 200 & 204.

On the other hand, SARPs and PANS regulate the technical performance requirements of GNSS in general,[200] GPS, and GLONASS respectively in great detail,[201] and this gives rise to a question: do GNSS providers in general, GPS providers, and GLONASS providers or their States have the legal obligation to comply with those technical codes? This answer depends on whether SARPs and PANS themselves have legal binding force directly on contracting States of the Chicago Convention and its Annexes; this has raised questions in courts, in literature and in practice.[202]

Although ICAO practices try to de-emphasise and blur the distinction between Standards, Recommended Practices and PANS,[203] their legal significance is not on the same plane. The reasons are as follows:

> *For one thing*, it is generally accepted that the legal force of PANS is unequal to, or lower than[204] that of SARPs.[205] PANS are not provided for in the Chicago Convention and do not come under the obligation imposed by Article 38 of the Chicago Convention to notify differences in the event of non-implementation.[206] That being said, the formulation of PANS requires the *approval* of the ICAO Council only with a simple majority vote,[207] while SARPs are *adopted* by the Council with a two-thirds majority vote.[208] However, the author holds that the status of PANS is higher than guidance materials such as the GNSS Manual and the GANP, as the latter are only for the purpose of instruction, whereas the former is much more for information on operational procedures connected to aviation safety. It should be stated that PANS do not have any binding force.
>
> *For another thing*, many commentators discuss the legal effect of SARPs as a whole, but do not make a distinction between *Standards* and *Recommended Practices*.[209] Against this background, opposing views exist: one school of thought insists that SARPs have the status of international treaties,[210] while another school of thought holds that they do not.[211] The author supports the latter. The first attempt at international standardisation and formulation of procedures of civil aviation was made in the 1919 Paris

200. *See* Section 3.7.2, Volume I, Annex 10 to the Chicago Convention.
201. *See* Section 3.7.3, Volume I, Annex 10 to the Chicago Convention.
202. *Cf.* Mendes de Leon, *supra* note 67, at 25.
203. Huang, *supra* note 136, at 182 & 191; Also *see* Abeyratne, *supra* note 98, at 83.
204. Ludwig Weber, *The Chicago Convention*, in Paul Stephen Dempsey & Ram S. Jakhu (eds) Routledge Handbook of Public Aviation Law (Routledge, 2017), at 101.
205. Milde, *supra* note 41, at 174-175.
206. ICAO, *Procedures for Air Navigation Services: ICAO Abbreviations and Codes*, Doc 8400, 2010, at viii.
207. Articles 52 & 90 of the Chicago Convention; Huang, *supra* note 136, at 58; Milde, *supra* note 41, at 175.
208. *See* Articles 37 & 90 of the Chicago Convention; von der Dunk, *supra* note 23, at 8.
209. The author does not share this approach and holds that the legal effect of *Standards* and that of *Recommended Practices* are different. *See* below for further discussion.
210. Bradford W. Parkinson & James J. Spilker, et al., *Progress in Astronautics and Aeronautics: Global Positioning System: Theory and Applications-Volume II* (American Institute of Aeronautics and Astronautics, Inc., 1996), at 334.
211. Mendes de Leon, *supra* note 67, at 24.

Convention through 8 Annexes,[212] and notably, those Annexes were given the same legal force as the Paris Convention itself.[213] However, the legal status of the Annexes to the Chicago Convention represents a clear departure from the previous practice under the Paris Convention.[214] According to Article 54 of the Chicago Convection which regulates the mandatory functions of the ICAO Council, SARPs are designated as Annexes to the Chicago Convention 'for convenience', and this wording indicates that SARPs are not integral parts of the Chicago Convention, but serve only as an easy way of codification.[215] Moreover, the *travaux preparatories* of the Chicago Convention clarified that "in fact as a necessary consequence of that flexibility, the Annexes are given no compulsory force. It remains open to any State to adopt its own regulations in accordance with its own necessities."[216] In academia, the following three reasons have been posited to deny the treaty status of SARPs: (i) all the procedures of adoption and update of SARPs were done through the ICAO Council rather than a Diplomatic Conference and national ratification, which are essential processes for concluding and amending a treaty;[217] (ii) States may choose to depart from international standards if they find it impracticable to comply with them,[218] and each State is the sole authority to determine what is 'practicable' and what is not;[219] and (iii) the Chicago Convention did not create any machinery or procedures for the 'enforcement' of compliance with SARPs,[220] and the implementation of SARPs merely relies on the moral obligation of States.[221]

However, the author holds that the fact that SARPs do not have treaty status does not mean that Standards (contrary to Recommended Practices) do not have any legal binding force.[222] There is a necessity to distinguish between the legal effect of Standards and that of Recommended Practices. According to Article 38 of the Chicago Convention, the legal obligation of immediate notification for non-compliance only connects with Standards, not Recommended Practices. Besides, the term 'shall' is used

---

212. Milde, *supra* note 41, at 167.
213. Article 39 of the Paris Convention, which reads as follows:

> The provisions of the present Convention are completed by the Annexes A to H, which, subject to Article 34 (c), shall have the same effect and shall come into force at the same time as the Convention itself.

214. Milde, *supra* note 41, at 168.
215. *Ibid.*; Abeyratne, *supra* note 98, at 103.
216. The US Department of State, *Proceedings of the International Civil Aviation Conference: Vol. 1* (United States Government Printing Office, 1948), at 92.
217. *Cf.* Milde, *supra* note 41, at 167; Huang, *supra* note 136, at 66; Eugene Sochor, *The Politics of International Aviation* (Macmillan Press, 1991), at 58.
218. *See* Mendes de Leon, *supra* note 67, at 24; Article 38 of the Chicago Convention.
219. Milde, *supra* note 41, at 170.
220. *Ibid.*, at 176.
221. Abeyratne, *supra* note 98, at 75.
222. Another viewpoint insists that SARPs, where Standards are included, are non-binding and lacks mandatory force. *See* Thomas Buergenthal, *Law-Making in the International Civil Aviation Organization* (Syracuse University Press, 1969), at 76.

in Standards, which typically indicates a mandatory sense which the drafters intend and the courts hold,[223] whereas a Recommended Practice containing the same elements as a Standard could only use the term 'should' instead.[224]

Therefore, different from Recommended Practices whose non-compliance does not constitute an infraction of the Chicago Convention,[225] Standards have compulsory legal force and they are binding on contracting States which do not file any differences from Standards pursuant to Article 38 of the Chicago Convention.[226] The Chicago Convention does not allow for a situation where States neither comply, nor file differences at the same time.[227] Professor Bin Cheng concluded the above phenomenon as the *indirect* legal force of Standards;[228] the author holds the same viewpoint but from another perspective, namely that Standards do not have *full* legal effect as treaties, but they do have *conditional* legal force. The *condition* thereof is the State's notification to ICAO for not complying with Standards. In other words, a contracting State of the Chicago Convention is entitled to not comply with Standards but only after the required notification has been sent accordingly; before that notification, Standards have a compulsory binding force which the Contracting State is obligated to comply with.[229]

A question may arise here as to whether a State which failed to notify its departure from, or non-compliance with Standards constitutes an international fault if damage is thus caused. Luckily, such a contingency has not yet been recorded in the history of ICAO,[230] but it is not only an abstract theoretical concern in the context of GNSS. Even though the US and Russian governments have pledged full cooperation with ICAO in the development of SARPs,[231] both GLONASS and GPS suffered well-known system failures for a period of more than ten hours in April 2014 and February 2016 respectively.[232] Given that the US and Russian governments did not notify the non-compliance of system parameters as required by Section 3.7.3.2, Volume I Radio Navigation Aids, Annex 10 Aeronautical Telecommunications to the Chicago Convention,[233] an air accident did unfortunately happen due to the system malfunction. In this

223. Bryan A. Garner (Eds), Black's Law Dictionary (WEST, 2009), at 1499.
224. ICAO, *supra* note 181, at 7.
225. Abeyratne, *supra* note 4, at 38.
226. Huang, *supra* note 136, at 60; von der Dunk, *supra* note 23, at 14; Henaku, *supra* note 32, at 36.
227. Assad Kotaite, *Sovereignty under Great Pressure to Accommodate the Growing Need for Global Cooperation*, 50(10) ICAO Journal 1995, at 20.
228. Bin Cheng, *Centrifugal Tendencies in Air Law*, 10 Current Legal Problems 1957, at 205.
229. Ronald I.C. Bartsch, *International Aviation Law: A Practical Guide* (Ashgate, 2012), at 59; Weber, *supra* note 204, at 21.
230. Milde, *supra* note 41, at 171.
231. Andrade, *supra* note 99, at 87.
232. ICAO, *The Current Status and Further Development of the GLONASS Orbital Grouping in Support of Multi-Constellation GNSS Implementation*, A39-WP/452, TE/199, 20/9/16, presented by the Russian Federation, at 3; Jonathan Amos, *Map illustrates 'Russian GPS' Failure*, http://www.bbc.com/news/science-environment-26957569, last accessed 5 February 2018; Chris Baraniuk, *GPS Error Caused '12 Hours of Problems' for Companies*, http://www.bbc.com/news/technology-35491962, last accessed 5 February 2018.
233. It was said that the GLONASS error in April 2014 was not communicated to the users. G. Bertler, et al., *GLONASS and Multi-GNSS in the IGS: Lessons learned from GLONASS Service Disruptions*, presented to 13th Meeting of the National Space-Based Positioning, Navigation, and Timing (PNT) Advisory Board, Washington DC, 3-4 June 2014, at 20.

hypothetical case, is the US or the Russian government liable for damage caused by the lack of immediate notification of non-compliance to ICAO, or should this government be?

Before solving the above questions, one more question arises: does that the very short time (around ten hours in both cases of GLONASS and GPS) of non-compliance fall within the scope of Article 38 of the Chicago Convention? It seems that scholars seldom touch upon this question.[234] The author holds that, for the safety of civil aviation in each single moment, the term *practice* used in Article 38 includes not only strategic rule-making, but also daily tactical operation. A State is legally responsible for the *immediate* notification to ICAO, as long as that State "finds it impracticable to comply in *all* respects" with Standards regardless of the length of non-compliance,[235] and ICAO should hold a repository of all such differences for evidence or whatever purposes.[236]

If a State failed to send the required notification, the predominant viewpoint regards it as a breach of Article 38 of the Chicago Convention.[237] However, does that breach qualify as a legal basis to claim compensation? One opinion is that "*a contracting State* would, therefore, be liable to *another contracting State* if the latter, or one of its nationals, suffers damage as the result of a mistaken belief, induced by the lack of notification, that the former contracting State was complying with a given international standard."[238] Another opinion is that "nor do Standards owe their efficacy to the right acquired by *some injured person* to damages against a *State* omitting to notify."[239] Although at first sight these two views seem contradictory, actually they are not: the former proceeds from the liability model of *State versus State*; while the latter proceeds from the model of *individual versus State*.

The Chicago Convention is inherently *public* (contrary to *private*) international air law,[240] and there is no doubt that this *public* nature extends to its Annexes as *technical* codes of conduct. What public international air law regulates is the relationship between States, rather than the issue of compensation under private law to individual bodies. If a contracting State caused damage due to the breach of Standards to another contracting State and/or its nationals, the former should be liable to 'another contracting State', rather than directly to 'some injured person'. More importantly, the obligation of notification for non-compliance with GNSS parameters regulated by Standards lies on a *State* as a regulator or a public authority, rather than as a signal provider under private law.[241]

Here it could be concluded that standards play a role in the context of GNSS through the following two mechanisms:

---

234. As of 13 April 2018, the author has not found enough literature on this matter.
235. Weber, *supra* note 76, at 305.
236. Bartsch, *supra* note 229, at 58.
237. Chia-Jui Cheng (Eds), *Selected Works of Bin Cheng* (Brill Nijhoff, 2018), at 52; Cheng, *supra* note 228, at 200; Weber, *supra* note 204, at 21.
238. Cheng (Eds), *ibid.*
239. Henaku, *supra* note 32, at 36.
240. Mendes de Leon, *supra* note 67, at 9; Milde, *supra* note 41, at 17; Dempsey, *supra* note 68, at 8.
241. For the difference between GNSS provider and GNSS regulator, and the distinction between GNSS administrative liability and GNSS civil liability, *see* §2.03[A] & §2.03[B] respectively.

(i) States which provide ANS based on GNSS are obligated to provide air navigation facilities *in accordance with SARPs*, which include the requirements of GNSS parameters. If damage is caused in the provision of ANS, the ATC liability or ANS liability will be triggered due to its violation of Article 28 of the Chicago Convention, and this will be discussed in §4.04.

(ii) GNSS provider States (which mainly refers to the US and Russia)[242] are obligated to comply with Standards (Sections 3.7.2 and 3.7.3, Volume I, Annex 10 to the Chicago Convention) before they send any notification of non-compliance to ICAO. If damage is caused because of either non-compliance or non-notification, the GNSS provider State should be responsible and liable for its breach (maladministration) of Article 38 of the Chicago Convention.[243]

[F] Brief Conclusion

ICAO has for decades provided a forum for the rule-making of GNSS in the context of civil aviation, consequently publishing a series of regulatory and guidance materials related to GNSS. The Policy Statement was issued in the 1990s to address the most fundamental principles of GNSS, but it is neither legally binding, nor relevant to the issue of civil liability. Although guidance materials such as the GNSS Manual and the GANP are continuously updated, they hardly touch upon legal elements of GNSS and are just for illustrative or descriptive purposes without any legal significance. The political exchanges of letters were considered as a starting point in moving further to an international convention on GNSS matters in civil aviation,[244] but no further agreements or treaties were concluded except for a Charter being adopted and recognised as a pre-legislation Assembly Resolution. The exchanged letters and the Charter however neither address the issue of civil liability directly, nor indicate the intention of relevant States to conclude a legally binding treaty. Standards (unlike Recommended Practices and PANS) have compulsory force before a contracting State sends its notification of non-compliance to ICAO, but a breach of Standards leads to either ATC civil liability, or administrative liability of the GNSS provider State as a regulator rather than a provider under private law, which are different with the regime of GNSS civil liability.

Therefore, the ICAO regime is currently inappropriate and too far from being adequate to deal with the issue of GNSS civil liability. This is based on the reasoning that: the purpose of ICAO relevant materials is mostly for public governance or technical assurance for the implementation of GNSS in the sense of public law; and at

242. This is because Annex 10 to the Chicago Convention only lays down the performance parameters of GPS and GLONASS, and those of Galileo, BDS, and other specific GNSSs are not included.

243. Another topic which could be further discussed in this part is whether SARPs, whose adoption or amendment requires a two-thirds majority of the Member States represented on the Council, may get the status of customary international law where non-compliance does not exist in national legislative practices.

244. ICAO, *supra* note 118, at 6.

the same time, none of those materials address legal rules of civil liability directly, even though some of them could be used as reference materials in court to justify the accountability of States providing either GNSS signals or ANS based on GNSS. The next two sections will draw attention to private international air law.

§4.04 ATC CIVIL LIABILITY FOR DAMAGE CAUSED BY GNSS?

[A] Arguments on the Relationship Between ATC Civil Liability and GNSS Civil Liability

ICAO held detailed and lengthy discussions concerning the possible approaches to deal with the legal implications of GNSS in general and the issue of civil liability in particular. Nevertheless, no coordinated political will and agreed specifics have been expressed as yet.[245] Although dissatisfaction with the *status quo* of the legal framework governing GNSS activities in civil aviation was expressed by most GNSS user States,[246] there has always been a loud voice, mainly presented by GNSS provider States, that it is unnecessary to establish a new universal liability system for GNSS because there is no indication that the current liability regime of ATC under domestic law is unable to cope with GNSS.[247] Accordingly, the question of whether the issue of GNSS civil liability can be addressed appropriately and adequately under the present ATC civil liability regime remains open in academia.

Many scholars hold that GNSS is technically no more than another navigational aid for civil aviation,[248] for which it should not be treated differently in any respect.[249] The advent of GNSS merely adds another potential ultimate cause of damage to those already in existence, such as traditional navigation errors, human negligence, engine failure or force majeure.[250] GNSS should be under the same liability provisions as all other navigation aids, so that there is no need for a fundamentally different interpretation or separate legal paradigm in addition to the present regime of ATC civil liability.

The author respects the fact that civil liability regimes for GNSS and ATC share many similar fundamental issues,[251] yet he would also like to address the significant

---

245. *Cf.* Milde, *supra* note 128, at 197.
246. ICAO 34, *supra* note 53, at 2.
247. *Ibid.*
248. Milde, *supra* note 128, at 211; Andrade, *supra* note 65, at 123.
249. Battama Kantasuk, *General Legal Issues Concerning GNSS and the Impact on Developing Countries* (McGill LLM Thesis, 1997), at 57.
250. von der Dunk, *supra* note 10, at 132.
251. As Professor Michael Milde summarised in his article about the fundamental issues of Liability of Air Traffic Control Agencies, which are the same issues for GNSS:

    (i) whether there is a need for an international instrument or model legislation or its determination of the choice of applicable law;
    (ii) whether any liability is to be based on proven fault, fault with a reversed burden of proof or whether there should be strict (absolute) liability;
    (iii) whether the liability is to be limited or unlimited;
    (iv) whether State organs providing ATC could claim jurisdictional immunity;
    (v) the courts of which State would have jurisdiction to consider claims, etc.

challenges raised by the introduction of GNSS in the implementation of CNS/ATM system.[252] In addition, it is also true that a lot of issues concerning the present ATC civil liability themselves remain unanswered,[253] therefore it has no capability to deal with the complicated issue of GNSS civil liability. To substantiate the author's viewpoint, it is first necessary to determine how big the difference between CNS/ATM systems based on GNSS and traditional ground-based ATC is before bringing forward any further legal analysis on this subject.

[B] GNSS-Based CNS/ATM Versus Traditional ATC

Safety is always the fundamental prerequisite of civil aviation,[254] for which the concept of ATC was developed in 1929 when the first professional Air Traffic Controller was hired.[255] The history of ATC has evolved generationally from visual tools such as coloured flags and light guns, to a system of radio navigation aids which is either ground-based or satellite-based.[256] Furthermore, the organisation of the provision of ATC service is moving from a government-run and tax-funded entity to a national, regional or international corporation funding derived from user fees among the most developed States.[257] The primary function of ATC service is to prevent a collision between aircraft or with obstructions, and to maintain an orderly flow of air traffic along with the goal of economy during all phases of flight.[258] Notably, ATC forms the core of Air Traffic Service (ATS) which is often called ANS;[259] yet, ATC service is almost invariably provided by the same agency that provides other kinds of ATS, such as altering service and flight information service.[260] For the purpose of this book, the author does not make a distinction between the terms ATC and ATS or ANS in the context of civil liability.

Due to an increasing need for ATSs as a result of the dramatic development of civil aviation in recent decades, the traditional ATC system was inherently no longer sufficient, despite continuous improvements. First, the conventional ground-based navigation aids are often inaccurate, in which case navigational errors have unfortunately caused or contributed to several disasters such as the Korean Air Flight 007 accident. This could have been prevented by better means of precise navigation and

---

See Milde, *supra* note 128, at 211.

252. John Mark MacKeigan, *Liability of Air Traffic Services Providers: The Impact of New Systems and Commercialization* (McGill LLM Thesis, 1996), at i.
253. *Cf.* Ingrid Lagarrigue, *ATC Liability and the Perspectives of the Global GNSS: Is An International Convention viable?* (McGill LLM Thesis, 1994), at 39.
254. Milde, *supra* note 128, at 196.
255. Michael S. Nolan, *Fundamentals of Air Traffic Control* (Cengage Learning, 2011), at 5.
256. *Ibid.*, at 4.
257. Rui Neiva, *A History of Air Traffic Control Provision in the United States*, https://www.enotrans.org/article/history-air-traffic-control-provision-united-states/, last accessed 7 March 2018
258. Doo Hwan Kim, *Some Considerations on the Liability of Air Traffic Control Agencies*, 10(1) The Korean Journal of Air & Space Law and Policy 1988, at 272; Chapter 1, Annex 11 to the Chicago Convention; Milde, *supra* note 128, at 196.
259. Stephan Hobe, Nicolai von Ruchteschell & David Heffernan (Eds), *Cologne Compendium on Air Law* (Carl Heymanns Verlag, 2013), at 681.
260. Chapter 1, Annex 11 to the Chicago Convention.

capable surveillance on malfunction.[261] Second, the traditional ATC system is not implemented on a uniform basis worldwide due to the expense,[262] and it is therefore composed of individualised networks often dependent on poorly coordinated national infrastructure, especially in developing countries.[263] Third, the ground-based ATC system has very limited geographical coverage either because of the inherent characteristics of the infrastructure, or the topography of the area.[264]

After recognising those limitations of the terrestrial-based system, ICAO responded to the need for a technical upgrade by implementing CNS/ATM systems.[265] Scholars have generally presented the significant benefits of CNS/ATM systems based on GNSS.[266] GNSS has been a critical technical component of the new approach of ATC since the introduction of CNS/ATM systems (*see* §4.02[B]), and it has very meaningfully enhanced the safety of air navigation.[267] However, no matter how advanced GNSS technology is, it does not go beyond being another form of navigation aid or technical means of ATC system in civil aviation.[268]

On the one hand, the benefit of GNSS is mainly through the implementation of CNS/ATM systems, but GNSS is not the whole picture of the concept of CNS/ATM systems which also includes communication and surveillance.[269] It is the entire CNS/ATM systems, rather than GNSS itself, which are supposed to replace the existing traditional national ATC mechanism.

On the other hand, GNSS indeed does provide the same service as ground-based navigation aids do to international civil aviation, albeit with superior and precise technology.[270] Nevertheless, what GNSS can provide is location-based data for pilots and ATC service providers, rather than the ATC service directly to pilots on-board. The primary application of GNSS in civil aviation is still under the concept or framework of ATC.

Therefore, the author agrees with the argument that GNSS is another form of navigation aid of ATC, even though the manner in which it differs from conventional ground-based navigation aids, including VOR/DME (VHF Omnidirectional Range/Distance Measuring Equipment), Microwave Landing System (MLS), and Inertial Navigation System (INS),[271] is also apparent. This difference is not substantial, but it does determine whether GNSS is too distinct to apply the present regime of ATC civil liability.

---

261. Milde, *supra* note 128, at 196.
262. MacKeigan, *supra* note 252, at 13.
263. Ranjana Kaul, *Liability Implications of the Use of Global Navigation Satellite Systems (GNSS) for Communication, Navigation, Surveillance/Air Traffic Management (GNS/ATM) in Civil Aviation: With Special Focus on India*, XXXV(I) Annals of Air and Space Law 2000, at 418.
264. Henaku, *supra* note 32, at xv.
265. ICAO, *supra* note 22, at 12-1; Kaul, *supra* note 263, at 430; Abeyratne, *supra* note 4, at 36.
266. Milde, *supra* note 128, at 196; Abeyratne, *supra* note 4, at 37.
267. Milde, *ibid.*; Kaul, *supra* note 263, at 414.
268. Kantasuk, *supra* note 249, at 57.
269. *See* Australian Government Civil Aviation Safety Authority, *CNS/ATM: Transforming Airspace Management* (Civil Aviation Safety Authority, 2017), at 8-9.
270. Kaul, *supra* note 263, at 430.
271. EUROCONTROL, *Conventional Navaids*, http://www.eurocontrol.int/articles/conventional-navaids, last accessed 8 March 2018.

First, unlike conventional navigation aids which can only provide limited *local* service coverage divided by Flight Information Regions (FIR) or physical territorial State boundaries,[272] GNSS is truly *global* as it can help facilitate more international seamless air navigation throughout unified airspace in all phases of flight worldwide.[273] The use of GNSS as a navigation aid can eliminate the need for a variety of ground systems operated by different States:[274] for example, one station under a GNSS-based CNS/ATM concept may cover the air traffic management (ATM) of the entire Pacific Ocean Region.[275] This may lead to the creation of more international or regional organisations like EUROCONTROL which provide ATC service internationally or regionally.

Second, conventional navigation aids are generally developed and operated by local ATC entities, thus each State keeps tight control over conventional navigation facilities constructed in its territory or jurisdiction.[276] As discussed in Chapter 1, however, most user States have to bear the monopoly of very few GNSSs which are neither constructed in their territory, nor operated by them. Therefore, GNSS as a navigation aid is out of the control of those States providing ATC service, and this may increase their concerns about risk of liability due to a malfunction of GNSS.[277]

Third, the impact of GNSS failures on ATC is more critical than that of failures of individual terrestrial navigation aids, and this due to the size of the airspace that could be affected, especially when GNSS is proposed to be used as a sole service in the future.[278] The proposal of GNSS as the sole means of air navigation offers the possibility for many States to dismantle some or all of their existing ground-based navigation.[279] The intensifying reliance on GNSS, as opposed to traditional, terrestrial-based navigation aids, increases the concerns of users and leads to a much stronger need to address GNSS civil liability.[280]

At this point, a brief conclusion can be made that the GNSS-based CNS/ATM system constitutes a fundamental change – although in essence it is not – in the technologies applied to traditional ATC.[281] The big difference made by the introduction of GNSS may disrupt the former balance and social reliance among those conflicting interests on the provision of ATC service since it brings new service providers into play with new interrelationships between them, and it raises new concerns regarding the issue of civil liability.[282] The following sections discuss whether Article 28 of the

272. Jiefang Huang, *Development of the Long-Term Legal Framework for the Global Navigation Satellite System*, XXII Annals of Air and Space Law 1997, at 590.
273. Abeyratne, *supra* note 4, at 37.
274. ICAO, *Global Navigation Satellite System (GNSS) Manual*, First Edition – 2005, Doc 9849, AN/457, at 1-2.
275. Huang, *supra* note 272.
276. Henaku, *supra* note 32, at xvii.
277. ICAO, *Operational Concept for Next Generation GNSS*, PBNICG/4 – IP/06, Agenda Item 11, 15/03/17, Presented by Secretariat, at 20; Milde, *supra* note 128, at 199.
278. EUROCONTROL, *EUROCONTROL Policy on GNSS for Navigation Applications in the Civil Aviation Domain*, Action Paper, SCG/8/AP10, 28/04/2008, at 11.
279. Weber, *supra* note 5, at 101.
280. *Ibid.*
281. MacKeigan, *supra* note 252, at 15; Murray, *supra* note 179, at 395.
282. *Ibid.*

Chicago Convention – the legal basis of the provision of ATC service – and the present regime of ATC civil liability are able to deal with those differences and challenges appropriately and adequately.

[C] Article 28 of the Chicago Convention

[1] State Responsibility for the Provision of ATC Service

The Chicago Convention, signed in 1944, established the core principles permitting international civil aviation.[283] The complete and exclusive sovereignty of a State over the airspace above its territory is confirmed by Article 1 (Sovereignty) as a cornerstone-rule of public international air law,[284] and Article 28 (Air navigation facilities and standard systems) pertaining to the responsibility of States for the provision of ANSs in their national airspace. Item (a) of Article 28 of the Chicago Convention is generally treated as the legal basis for the provision of ATC service,[285] which reads as follows:

> Each contracting State undertakes, so far as it may find practicable, to:
>
> (a) Provide, in its territory, airports, radio services, meteorological services and other air navigation facilities to facilitate international air navigation, in accordance with the standards and practices recommended or established from time to time, pursuant to this Convention [...].

While this provision urges contracting States to provide air navigation facilities, it indeed leaves them room by including the phrase 'so far as it may find practicable'.[286] Actually, the provision of ATC service is a sovereign task,[287] thus States are accorded the discretion to both determine the entity that is, by a governmental agency, a commercialised or privatised entity, or foreign or international organisations, etc., and establish air navigational rules and facilities in their territories.[288] Nevertheless, it is also the responsibility of each State to ensure the highest practicable degree of uniformity in all matters concerned with the safety of air navigation through compliance with SARPs, particularly those termed Standards, before the notification of non-compliance was sent to ICAO (*see* §4.03[E]).[289]

Although the above terms are seldom controversial, the question that is much more critical for this book is: what obligations and when do States have to incorporate such new technology as GNSS in their ATC facilities entailing their responsibility for such services?

---

283. ICAO, *The History of ICAO and the Chicago Convention*, https://www.icao.int/about-icao/History/Pages/default.aspx, last accessed 10 March 2018.
284. Huang, *supra* note 272; Weber, *supra* note 5, at 100.
285. Kaul, *supra* note 263, at 425.
286. Murray, *supra* note 179, at 390.
287. Hobe, von Ruchteschell & Heffernan (ed.), *supra* note 259, at 683.
288. *Cf.* Abeyratne, *supra* note 4, at 37.
289. UNIDROIT, *supra* note 46, at 24.

[2] *Do States Have to Introduce GNSS under Article 28 of the Chicago Convention?*

Since the very beginning of the concept of CNS/ATM systems, a gesture has been made very clearly by ICAO that the implementation of either CNS/ATM systems as a whole, or GNSS in particular is compatible with Article 28 of the Chicago Convention and leaves unaffected the responsibility of States for provision of ANSs under that provision.[290]

Furthermore, certain scholars treated Article 28 of the Chicago Convention as the *legal basis* for the implementation of GNSS for CNS/ ATM systems.[291] That said, the author holds that Article 28 of the Chicago Convention does not offer States much legal obligations with respect to the introduction of GNSS as a navigation aid within their respective airspace. Hence, the use of GNSS is not a compulsory obligation on all States, at least not currently. The author presents the following arguments as support for this statement:

(i) the use of GNSS was not even foreseen in the 1940s when the Chicago Convention was drafted. Article 28 of the Chicago Convention is thus 'neutral' in this regard. Therefore, under the provisions of Article 28, no State is historically obligated to make use of such space technology as long as other practicable technical means are available for its ATC service.[292]

(ii) due consideration should also be given to the principle of complete and exclusive sovereignty of States over the airspace above their territory - an essential part of customary international air law enshrined in the 1919 Paris Convention and the 1944 Chicago Convention[293] - because most ICAO documents concerning GNSS expressed that it was not their intention to limit that principle (*see* §4.03).[294] Indeed, the responsibility of the provision of ATC service derives from the exclusive sovereignty over the airspace of each State. As such, the use of GNSS may not be imposed on States against their will, and their support of GNSS depends entirely on their sovereign political will, considering that, technically speaking, the introduction of GNSS is bound to infringe on States' sovereignty.[295] The acceptance of GNSS by non-provider States as their fundamental infrastructure requires substantial concessions from that sovereign State.[296] Although certain ICAO policies and guidance documents reiterate, often from a public 'law' angle, that the implementation of GNSS may neither infringe, nor impose restrictions on State sovereignty, authority, or responsibility in the control of air navigation

290. ICAO 75, *supra* note 53, at 2; ICAO, *supra* note 22, at 12-1.
291. Kaul, *supra* note 263, at 428 & 429.
292. Milde, *supra* note 128, at 200.
293. Prachee Kulkarni & Pablo Mendes de Leon, *Liability for Damage in Civil Aviation Within the Context of GNSS*, 3(1) Outer Space Committee Newsletter 2000, at 29.
294. Andrade, *supra* note 99, at 87.
295. Milde, *supra* note 128, at 198; Kotaite, *supra* note 227, at 21.
296. Milde, *ibid.*

and the promulgation and enforcement of safety regulations,[297] again, however, these materials show no legal binding force which would not actually reduce the concerns of non-provider States (*see* §4.03).[298] A political or legal arrangement, such as the conclusion of a series of agreements (*see* §5.03[C][2]), is thus needed to rebalance the interests between provider and non-provider States.

(iii) what Article 28 of the Chicago Convention requires is the provision of 'air navigation facilities', but it does not specify what kind of air navigation aids should be used. As already mentioned, GNSS is only one form of air navigation aid and will not always present the latest technical means of ATC service as the development of technology.

(iv) any ATC facilities that a State needs to provide are only those which a State 'may find practicable'. At present, the provision of traditional ground-based infrastructure may be required as compulsory, but the use of GNSS does not yet seem practicable for most States. Even though GNSS is made operational in each corner of the globe, States may still insist that GNSS controlled or operated out of their territory is impracticable and not available for use in their sovereign airspaces. The Chicago Convention does not offer a clear mechanism to define the term 'practicable'. Therefore, the introduction of GNSS could qualify as 'desirable' much more than 'necessary', at least before GNSS is popularly used as a navigation aid in the majority of States.

In brief, Article 28 of the Chicago Convention is highly significant concerning the provision of ATC facilities as a whole, but nothing in this provision explicitly or even implicitly addresses GNSS itself.[299] The ultimate responsibility of States under Article 28 of the Chicago Convention for the provision of ATC facilities within their national airspaces would remain intact, regardless of whether GNSS is introduced.[300] In other words, unlike air navigation facilities as a whole, for the time being Article 28 of the Chicago Convention does not enforce any legal obligation on States to introduce GNSS.

[3] The Legal Implications of Article 28 on ATC Civil Liability and GNSS Civil Liability

Article 28 of the Chicago Convention is frequently read with the regime of ATC civil liability.[301] However, what Article 28 of the Chicago Convention establishes is clearly State *responsibility* concerning the provision and supervision of ATC facilities. Although, by logical extension, breach of this State responsibility will trigger State

297. Kulkarni & Mendes de Leon, *supra* note 293, at 29.
298. *See* Milde, *supra* note 128, at 198.
299. *See* Murray, *supra* note 179, at 391.
300. *Cf.* von der Dunk, *supra* note 23, at 24.
301. *See* Francis P. Schubert, *Warsaw Claims and ATC Liability: Addressing the Global Dimension of Aviation Liability*, XXXII(I) Annals of Air and Space Law 1997, at 252.

liability *under general international law,*[302] *responsibility* under Article 28 of the Chicago Convention should not be construed as *liability* (*see* §2.03[A]).[303] Article 28 of the Chicago Convention is actually silent about civil liability arising out of the failure of an ANS provider to fulfil its obligations to the ANS user.[304]

Strictly speaking, the State responsibility obligated by Article 28 of the Chicago Convention is to provide air navigation facilities, but it does not mention *how* those facilities are to be provided. Although there is a viewpoint which furthers that responsibility to ultimate liability of a State in the provision of ATC service,[305] regardless of whether the provider is a governmental body or not,[306] the author holds that what Article 28 of the Chicago Convention imposes on States is merely a general obligation for the construction of infrastructures for the use of civil aviation, and it is irrelevant to the issue of civil liability under private law. The similarities can be found in the civil liability regime of airports which has been fully commercialised in most States. In addition to air navigation facilities, a State is also obligated to provide airports under Article 28 of the Chicago Convention, but the civil liability for damage caused by the operation of airports is seldom claimed against that State by the theory of State liability based on State responsibility. In this regard, under the regime of ATC liability, a distinction should be made between State liability for its maladministration and State liability for the failure of ATC service, and the latter is actually 'ATC civil liability' under private law.

In addition, Article 28 of the Chicago Convention, as the fundamental instrument of *public* international air law, regulates only the relationship between sovereign States,[307] and it neither expressly prescribes any sanctions for breach of the undertaking by States to provide air navigation facilities, nor directly gives a cause of action for compensation to private parties to claim compensation for damage.[308] Unlike international space law, where the Outer Space Treaty separately regulates State responsibility and liability on the launch of a space object by Article VI and Article VII respectively, and where more importantly the latter clearly addresses the issue of civil liability by the words "[...] is internationally liable for damage to another State Party to the Treaty or to its natural or juridical persons [...]," Article 28 of the Chicago Convention does not mention any terms like 'liable', 'liability', 'damage' or 'compensation'.

In brief, the Chicago Convention is not a liability-based regime.[309] Article 28 of the Chicago Convention refers only to the responsibility for the provision of ATC facilities in accordance with ICAO SARPs, and it has no direct relevance whatsoever to

302. Yaw Out Mankata Nyampong, *Privatization and/or Commercialization of Air Navigation Services and Its Liability Implications*, presented to the International Conference on Contemporary Issues in Air Transport, Air Law and Regulation, New Delhi, India, April 23-25 2008.
303. ICAO 75, *supra* note 53, at A-4.
304. Kaul, *supra* note 263, at 426.
305. Francis Schubert, *The Liability of Air Traffic Control Agencies – The Ueberlingen Midair Collision Case Study*, presented to the International Institute of Air & Space Law, Leiden University, 11 October 2017, at 71.
306. Schubert, *supra* note 301, at 241.
307. ICAO 75, *supra* note 53, at A-4; Andrade, *supra* note 65, at 124.
308. Nyampong, *supra* note 302.
309. Abeyratne, *supra* note 37, at 24.

the civil liability issue of either ATC or GNSS. Thus Article 28 of the Chicago Convention does not serve as a legal basis of GNSS civil liability under private law.

[D] Why Is the Present Regime of ATC Civil Liability Not Appropriate for the Issue of GNSS Civil Liability?

[1] Challenges Arising from GNSS for the ATC Civil Liability Regime

Although it is generally accepted that legal issues including civil liability will and should not be a roadblock obstructing the implementation of a GNSS-based CNS/ATM concept, regulatory advancements in the field of air navigation could not be left too far behind.[310] The use of GNSS as a navigation aid marks a fundamental technological innovation compared with conventional ground-based facilities (*see* §4.04[B]), and this unavoidably brings with it a new set of challenges to the present legal regime of ATC civil liability.[311]

In the traditional picture of legal relationships concerning the provision of ATC service, the issue of civil liability used to focus on the ATC service provider, as the provider is usually the operator or the owner of conventional ground-based air navigation facilities. The insertion of GNSS provider actually creates new social relationships and conflicts of social interest in the chain of events which leads to an air accident, and this may lead to the need for new regulations.[312] For example, there is a clear distinction between GNSS provider and the entity providing value-added service based on GNSS for air navigation, that is the ATC service provider.[313] In this sense, for damage caused by CNS/ATM systems, a GNSS provider would likely be included in the lawsuits in order to increase the chance of full compensation claimed by victims.[314]

This notwithstanding, the role of GNSS in the context of civil aviation is merely another technical means of navigation aid (*see* §4.04[B]), and this urges us to carefully examine whether the present regime of ATC civil liability is adequate or applicable to the issue of GNSS civil liability before any new regime were specially considered for damage caused by GNSS in civil aviation. For that purpose, it is first necessary to understand how the law relating to ATC civil liability developed and where it now stands.[315]

[2] The Failure of Practice on the Proposal for a Multilateral Convention

GNSS, in most cases as a global system operated by foreign States, makes the provision of ATC service much more international due to not only the broader scope of service

310. Lazar Vrbaski, *Liability of Air Navigation Services Providers: Towards an International Solution*, 38(1) Air & Space Law 2013, at 34.
311. MacKeigan, *supra* note 252, at i.
312. Schubert, *supra* note 301, at 237.
313. Andrade, *supra* note 99, at 85; von der Dunk, *supra* note 10, at 144.
314. *Cf.* Schubert, *supra* note 301, at 237.
315. MacKeigan, *supra* note 252, at 15.

but also the foreign actors involved (*see* §4.04[B]). Therefore, it is reasonable to expect an international solution on ATC civil liability to take care of the above features raised by the introduction of GNSS.

Unlike the regime of air carrier civil liability which is regulated by several international conventions such as the Warsaw System and the Montreal Convention (*see* §4.05), ATC civil liability remains untouched by the majority of international instruments of private air law. Throughout its history, several drafts for a multilateral convention governing ATC civil liability were prepared in the lengthy discussions.[316] Unfortunately, however, ATC civil liability is such a complicated long-term issue that a consensus on this subject has not yet been reached in spite of decades of deliberations under the umbrella of ICAO,[317] and this situation may not change in the foreseeable future.

Nevertheless, if the agenda for the conclusion of a convention for ATC civil liability fortunately is somehow resumed, special attention would have to be addressed to the use of GNSS for the provision of ATC service. Notably, the question whether and how a GNSS provider is obligated to undertake civil liability if damage is caused by GNSS malfunction would have to be answered in that convention, so that a legal link between GNSS provider and ATC service provider or victims could be established (*see* §4.04[D][5]).

At the present stage, cases concerning ATC civil liability have no choice but to rely on the domestic law applicable according to the rules of conflict of laws.[318] Yet, the national regime of ATC civil liability has inherent limitations when it deals with damage caused by GNSS,[319] and this is discussed in the following sections.

[3] Fragmentation and Diversification of National ATC Civil Liability Regimes

The international movement of civil aircraft magnifies the effect of air accidents caused by GNSS-based CNS/ATM systems, and this indicates that damage may happen anywhere, victims may be from many different States, and claims may be lodged in

316. For example, a preliminary draft was presented by the Delegation of Argentina to the twenty-fifth Session of the Legal Committee of ICAO in 1983, and it was resubmitted several times after that; in addition, the International Federation of Air Traffic Controllers' Associations (IFATCA) had prepared its own draft Convention in 1976. *See* Kim Doo Hwan, *Some Considerations on the Liability of Air Traffic Control Agencies*, XIII(6) Air & Space Law 1988, at 268; Walter Schwenk & Rüdiger Schwenk, *Aspects of International Co-operation in Air Traffic Management* (Martinus Nijhoff Publishers, 1998), at 295; MacKeigan, *supra* note 252, at 47; Helene Sasseville, *Air Traffic Control Agencies: Fault Liability vs. Strict Liability*, X Annals of Air & Space Law 1985, at 242.
317. Kantasuk, *supra* note 249, at 45; MacKeigan, *supra* note 252, at 44 & 48; Lagarrigue, *supra* note 253, at 7; Schubert, *supra* note 301, at 241; Andrade, *supra* note 65, at 122; Kotaite, *supra* note 91, at 203.
318. Kulkarni & Mendes de Leon, *supra* note 293, at 29; Lagarrigue, *supra* note 253, at 2; Schubert, *supra* note 301, at 241.
319. Ingo Baumann, *State of Play in the European Union: Liability for GNSS Signals and Services*, (November/December Inside GNSS 2015), at 40.

jurisdictions all over the world.[320] Although ICAO has been making a great contribution to the uniformity of technical matters of air navigation through the implementation of SARPs, legal systems for ATC civil liability in different States remain very divergent.[321] Procedural rules and, in particular, rules relating to jurisdiction are nevertheless not fully adequate to bring all parties before the same court with a view to ensuring prompt and equitable compensation.[322] The fragmentation and diversification of national ATC civil liability regimes lead to uncertainty, therefore it is arguable that the issue of civil liability concerning GNSS-based CNS/ATM systems should be dealt with under a universal regime and should not be left to national law.[323]

In addition, most of the national regimes of ATC civil liability are not 'tailor-made' to reflect the specific characteristics of the domain,[324] but treat the ATC service provider as a governmental agency and thus apply the general regime of State liability.[325] Furthermore, it is reasonable to question the adequacy of national law in dealing with damage caused by international organisations providing ATC service such as EUROCONTROL.

[4] Immunity of States Providing ATC Service

Although a tendency of commercialisation and corporation has been raised for a long time,[326] the provision of ATC service remains a sovereign task and public function implemented by a governmental agency in the majority of States.[327] In most cases, the issue of civil liability for damage caused by the failure of ATC service is actually treated as a question of State liability under public law.[328] In this sense, non-provider States are reluctant to approve GNSS as their navigation aid because GNSS, as an element outside of State control, may increase the risk of State liability.[329]

In the field of civil aviation, the place of an accident is of great importance to deciding the law applicable and the court seizing the case,[330] but the place of an accident is more likely to be outside the jurisdiction of the State providing ATC service, which gives rise to multiple cross-border litigation. In most cases, the claim for State liability usually triggers the application of the doctrine of sovereign immunity and related principles, and this renders court action directly against a foreign governmental

320. *Ibid.*
321. ICAO 34, *supra* note 53, at 2.
322. *Ibid.*
323. *Ibid.*
324. Schubert, *supra* note 301, at 241.
325. *Ibid.*
326. Stephan Hobe, Katharina Irmen & Chiristian Plingen: *Privatization of German and Other European Air Navigation Service Providers and the Single European Sky Regulations*, XXXII(3) Air & Space Law 2007, at 168 & 177.
327. MacKeigan, *supra* note 252, at 17; Schubert, *supra* note 301, at 241; Kulkarni & Mendes de Leon, *supra* note 293, at 29.
328. MacKeigan, *ibid.*; Schubert, *ibid.*
329. ICAO, *supra* note 277, at 20.
330. Lagarrigue, *supra* note 253, at 8.

agency providing ATC service or GNSS signals quite difficult or impossible.[331] Therefore, the principle of sovereign immunity is a thorny problem for victims to get adequate and fair compensation through a juridical mechanism in those cases where the failure of ATC service is caused by the malfunction of GNSS.

A State intentionally relinquishing immunity in either an international convention, or a bilateral agreement is considered a way to help victims overcome the obstacle of the principle of sovereign immunity to claim against a governmental provider of ATC service,[332] but history demonstrates that it is indeed quite impossible (*see* §4.04[D][2]). In the author's view, a more practical way is to promote the commercialisation and corporatisation of ATC providers. Actually, like the context of GNSS (*see* §2.03[A]), States which provide ATC service usually have dual roles: one is to provide ATC service under private law, and the other is concerned with a regulatory oversight function.[333] There is currently a trend to separate the role of ATC service provider from that of regulator.[334] For that purpose, legal entities under private law, such as NAV CANADA,[335] are usually established as the provider of ATC service. Those commercialised and corporate providers of ATC service are mostly non-sovereign entities to which the principle of sovereign immunity does not apply, and they can generally be sued anywhere they have a presence.[336]

[5] Missing Link Between ATC Service Provider/Victims and GNSS Provider

A fair system of civil liability should guarantee victims an adequate compensation, and at the same time channel civil liability to an appropriate place,[337] that is the party who actually caused that damage. The insurer is not considered here. The provision of ATC service based on GNSS is a complex system where more than one party is involved. Any deficiency in the process of provision for any reason or in any manner whatsoever may result in an incalculable financial loss.[338] Similar to traditional air navigation facilities, in the provision of ATC service, GNSS may also become a contributing factor or even a sole cause of damage to aircraft, persons or goods on-board, or on the ground.[339]

331. ICAO 34, *supra* note 53, at 2.
332. Paul B. Larsen, *Air Traffic Control: A Recommendation for a Proof of Fault System without a Limitation on Liability*, 32 Journal of Air Law and Commerce 1966, at 5.
333. *See* Kotaite, *supra* note 91, at 203.
334. ICAO, *Rationalization of the ICAO SARPs System*, A36-WP/246, TE/80, 18/9/07, presented by the Civil Air Navigation Services Organisation (CANSO); ICAO, *supra* note 11, at Appendix B.
335. NAV CANADA is a private company that owns and operates Canada's civil air navigation service for aviation users in 18 million square kilometres of Canadian and oceanic airspace. NAV CANADA, *About Us*, http://www.navcanada.ca/EN/about-us/Pages/default.aspx, lase accessed 21 March 2018.
336. ICAO 34, *supra* note 53, at 2.
337. Kulkarni & Mendes de Leon, *supra* note 293, at 29.
338. Kaul, *supra* note 263, at 430.
339. ICAO, *supra* note 28, at 1-11-4.

[a] Fault-Based System of ATC Civil Liability

A review of the national law of ICAO Member States presenting different legal systems shows that the substantive law governing ATC civil liability is fault-based where negligence, that is a wrongful action or omission, must be proven.[340] It was argued that this national system of ATC civil liability is "reasonably adequate to determine or apportion liability arising from accidents involving failure or malfunction of GNSS systems."[341] The author nevertheless holds another view that under the national system of fault-based ATC civil liability, victims run the risk of inadequate compensation or no compensation at all in cases where damage was caused by defective GNSS signals. This is because usually the fault made by a GNSS provider cannot be contributed to an ATC service provider (*see* below).

Since the implementation of CNS/ATM systems, increasing automation places ATC service providers in a situation where they may no longer be able to deal with all of the flights under their responsibility at the time of a failure.[342] Damage caused by traditional ATC service is, in most cases such as the 2002 Überlingen mid-air collision, mainly related to *human error*,[343] but GNSS has a higher possibility of damage arising from a *technical failure*.[344] For example, defective GNSS signals may lead to incorrect information concerning air traffic being displayed, without the air traffic controller being aware of such aberration.[345] If damage was caused by this aberration, and provided the ATC service provider proves that it took every possible measure to avoid that damage or that it was impossible to take such measures, the ATC service provider is actually entitled to be exempted from civil liability.[346]

It may be arguable that an ATC service provider has the fundamental responsibility for the good performance of the technical infrastructure it uses through any technical tools or supervisory measures,[347] a breakdown of which cannot constitute a defence leading to exoneration of civil liability.[348] The author admits that it is within the freedom of States to decide whether or not to introduce GNSS as a navigation aid (*see* §4.04[C][2]), even though he would also like to address the fact that the operation of GNSS is actually beyond the *direct* control of the majority of States.[349] In the provision of GNSS-based ATC service, the role of ATC service provider is merely the body that transmitted those defective GNSS signals.[350] Without an advance notification, the specific air traffic controller is incapable of taking effective action in the

340. ICAO 34, *supra* note 53, at 2.
341. *Ibid.*
342. Schubert, *supra* note 301, at 255.
343. German Federal Bureau of Aircraft Accidents Investigation, *supra* note 82, at 81.
344. Andrade, *supra* note 65, at 140.
345. Schubert, *supra* note 301, at 255.
346. *Cf.* Article 7 of the *Preliminary Draft – International Convention on the Liability of Air Traffic Control Agencies*, submitted by the Republic of Argentina to the 25th Session of the ICAO Legal Committee (Montreal, 12-27 April 1983).
347. Abeyratne, *supra* note 4, at 46.
348. Hwan, *supra* note 316, at 271.
349. ICAO 34, *supra* note 53, at 2; ICAO 75, *supra* note 53, at 2; UNIDROIT, *supra* note 46, at 24; Kaul, *supra* note 263, at 438.
350. Schubert, *supra* note 301, at 255.

unscheduled interruption or defection of GNSS signals, particularly when GNSS is used as a sole means of navigation aid.[351] In this case, it is not the activity of the ATC service provider, but the GNSS provider who presents a risk,[352] so that it is hard to hold the former liable for the negligence required by the national fault-based system of ATC civil liability.

Nevertheless, can victims claim compensation directly against the GNSS provider which provides defective GNSS signals under a national fault-based system of ATC civil liability? The answer is unfortunately also in the negative. If victims try to sue the GNSS provider for breach of contract or breach of a duty of care in *general tort law* or *product law*,[353] that legal cause is not the one provided by the regime of *ATC civil liability*, but by the said *general tort law* or *product law*. Unless GNSS is specified in the national law for ATC civil liability, the legal link is missing between victims and the GNSS provider offering defective signals under the fault-based system of ATC civil liability (*see* Figure 4.1).

Figure 4.1 The Missing Link under the Fault-Base System of ATC Civil Liability

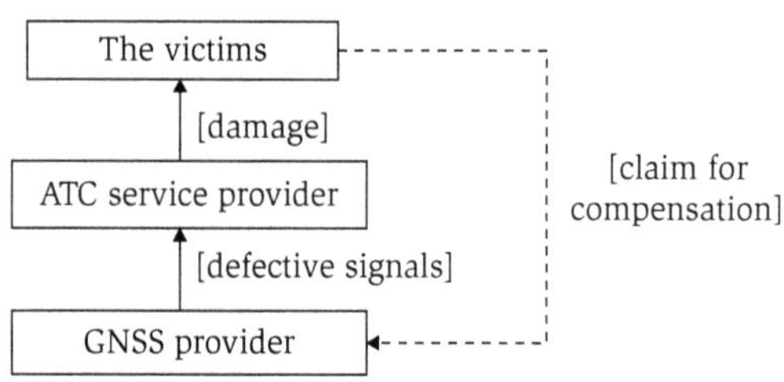

[b] No-Fault System of ATC Civil Liability

With the awareness of the difficulty in demonstrating negligence on the part of the ATC service provider, one viewpoint holds that the application of fault-based liability to damage resulting from a technical failure is in principle unacceptable.[354] Even though some States may therefore adopt the no-fault system of ATC civil liability, it is still however inadequate to deal with the issue of civil liability for damage caused by GNSS in the provision of ATC service. As discussed above, GNSS may be the sole cause of damage, but under the no-fault system of civil liability the ATC service provider is obligated to pay compensation for damage which it did not cause. Furthermore, in most cases it is others that provide defective GNSS signals causing damage. It is unfair to make an ATC service provider responsible and liable for what is completely out of its control. The risk of undertaking civil liability for the wrong action of others is actually

351. A few exceptions exist in the cases of GNSS providers such as the US, Russian Federation, the EU, Japan, India, and China.
352. Schubert, *supra* note 301, at 255.
353. Murray, *supra* note 179, at 396.
354. Jean-Pierre Bloch, *La responsabilité des services de la Circulation Aérienne* (Lausanne: Thonney-Dupraz, 1973), at 178, quoted from Schubert, *supra* note 301, at 255.

one of the main concerns of developing States in terms of whether to accept GNSS as their navigation aids before they are entitled to any remedies.

To remove that concern, a mechanism of a recourse action is needed under the no-fault system of ATC civil liability, as it may help channel civil liability from an ATC service provider, which relies on GNSS signals provided by others, to the GNSS provider who is indeed the real wrongdoer. In this regard, a link between GNSS provider and ATC service provider or the States having jurisdiction to provide ATC service under Article 28 of the Chicago Convention is required as the legal basis to make this recourse action possible; yet, that link is actually missing under the present regime of ATC civil liability (*see* Figure 4.2).[355]

Figure 4.2 The Missing Link under the No-Fault System of ATC Civil Liability

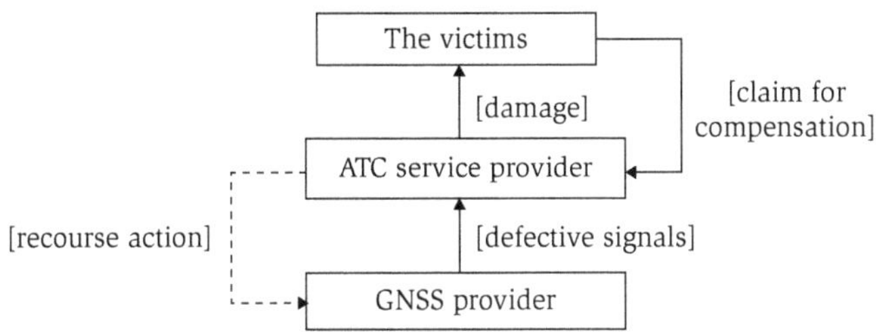

[c] The Roadmap to Establish the Missing Links

Generally speaking, a person claiming damages could sue another party either for breach of contractual obligations or for breach of a legal duty of care in tort law (product law is included).[356] Accordingly, a legal link can be established either by an agreement or by a legal norm.

For the missing link in Figure 4.1, the conclusion of an agreement seems impossible because GNSS signals are not provided to victims that are the passengers on-board and the people on the ground, but to an ATC service provider. Therefore, specific statutory provisions in national legislation or an international convention, which obligate a GNSS provider to be liable for damage caused by GNSS in the provision of ATC service, may better serve as the legal basis for the claim raised by victims against the liable GNSS provider.

For the missing link in Figure 4.2, the legal basis for the ATC service provider – who has paid compensation to victims because of damage caused by defective GNSS signals – to bring a recourse action against the GNSS provider who offered those defective signals can be served by both the statutory provisions mentioned above, and contract terms which determine how to delegate responsibility and allocate civil liability between an ATC service provider and GNSS provider.[357] Current CNS/ATM

355. *See* Kaul, *supra* note 263, at 438; ICAO 34, *supra* note 53, at 2; ICAO 75, *supra* note 53, at 2.
356. Murray, *supra* note 179, at 396.
357. *Cf.* Kotaite, *supra* note 91, at 204.

systems are nonetheless actually based on the *open signals* which are openly accessible,[358] thus no agreement on the use of GNSS signals has been established in practice.[359] Yet, this situation is a potential barrier to moving the regime of GNSS civil liability forward (*see* §5.02[C]).

Article 28 of the Chicago Convention does not prohibit a contracting State from delegating the responsibility of the provision of ATC service to other States, international organisations, or corporatised or private undertakings, therefore in practice, the cross-border provision formalised by way of cross-border arrangements is not rare.[360] The case of the 2002 Überlingen mid-air collision constitutes a typical example.

Inspired by the fact above, could we say that the use of GNSS signals provided by a foreign GNSS provider in the provision of ATC service actually constitutes a delegation of responsibility of providing ATC service?[361] The author would answer no. GNSS is used merely as a navigation aid in the provision of ATC service, and navigation is not the whole concept of CNS/ATM systems, in addition to which communication and surveillance are also essential to making ATC service possible (*see* §4.04[B]). Furthermore, a GNSS provider does not provide ANSs directly to airspace users, but this is done through the channel of ATC service provider. The delegation of the responsibility for providing ATC service usually refers to the delegation of actual provision of ANSs as a whole – unlike the part of navigation – in all/portions of national airspace to another 'craft brother' which provides the same or similar service.[362] This notwithstanding, GNSS is not the same as ANS under Article 28 of the Chicago

358. The US and Russian governments committed to provide GPS SPS and a standard-accuracy GLONASS channel to the world civil aviation community respectively; the EU provides the EGNOS Safety of Life Service to support civil aviation operations, and this kind of service is openly accessible. *See* respectively: ICAO 75, *supra* note 53, at Attachment B & C; European GNSS Agency, *ABOUT SoL*, https://egnos-user-support.essp-sas.eu/new_egnos_ops/services/about-sol, last accessed 29 March 2018.
359. For the purpose of this book, the author holds that the term GNSS is composed of core GNSS and augmented systems (*see* §2.02). But in practice, the allocation of civil liability should also be done between a core GNSS provider and an augmented service provider and therefore a legal link must also be established similar to that proposed in this section. Actually, an agreement or arrangement has been proposed between the US and India on the provision of GPS signals for augmentation by GAGAN. *See* Kaul, *supra* note 263, at 438.
360. Niels van Antwerpen, *Cross-Border Provision of Air Navigation Services with Specific Reference to Europe: Safeguarding Transparent Lines of Responsibility and Liability* (Leiden PhD Thesis, 2007), at 107.
361. ICAO has ever mentioned similar intention through the following words, but it did not open a further discussion on this specific topic:

> The implementation of GNSS leaves unaffected the responsibility of States under Article 28 for provision of air navigation services within their respective airspace. In fulfilment of such responsibility, certain issues relating to certification and authorisation of the use of GNSS, as well as the delegation of responsibility (emphasis added), will have to be resolved by the relevant States.

ICAO 75, *supra* note 53, at A-11.
362. *Cf.* Van Antwerpen, *supra* note 360, at 86.

Convention.[363] A GNSS provider is more like a *supplier* for navigation signals. Notably, the delegation of the responsibility for providing ATC service is usually based on a consensus in the form of an agreement, but such consensus or agreement is currently missing between an ATC service provider (including the State) and GNSS provider. Therefore, the model of civil liability for damage caused by GNSS in the provision of ATC service is different from the theory of cross-border provision of ATC service or the delegation of the responsibility for providing ATC service, where the delegator retains full responsibility for the execution of those services provided on their behalf to third parties.[364] In other words, an ATC service provider does not have to retain full responsibility for any consequence caused by GNSS, and accordingly, GNSS civil liability may be independent of the regime of ATC civil liability.

[E] Brief Conclusion

Although GNSS is in essence no more than another navigation aid, its introduction makes CNS/ATM systems a fundamental change from traditional terrestrial-based ATC system, and this creates certain challenges to apply the present regime of ATC civil liability for damage caused by GNSS. Article 28 of the Chicago Convention determines the responsibility of the provision of ATC facilities in accordance with ICAO SARPs, but it neither enforces any legal obligation on States about the introduction of GNSS for the time being, nor serves as the legal basis of ATC or GNSS civil liability under private law. Due to the failure of practice on the proposal for a convention, the regime of ATC civil liability has to rely on the domestic law applicable according to the rules of conflict of laws. Nevertheless, the national regime of ATC civil liability is not only too fragmented and divergent to ensure equitable compensation for all victims, but it may also place the obstacle of the principle of sovereign immunity before the claim against a governmental provider of ATC service.

Under the fault-based system of ATC civil liability, in a given case where damage was caused by defective GNSS signals and no negligence can be attributed to the ATC service provider, victims run the risk of inadequate compensation or no compensation at all. Under the no-fault system of ATC civil liability, it is unfair to make an ATC service provider responsible and liable for damage caused by GNSS which is completely out of its control. To make up for the inadequacy of a national regime for ATC civil liability due to the introduction of GNSS, the legal link between ATC service provider/victims and GNSS provider would need to be established either by an agreement, or by a legal norm (*see* §5.03[C][2]).

363. Gerhard Berz, *Authorisation and Operation of GNSS Aviation Services in Non-Core Constellation States*, presented to Civil GPS Service Interface Committee (CGSIC), Tampa, USA, 8 September 2014, at 5.
364. Case Number 4 O 234/05 H (Fourth Chamber), quoted from Mendes de Leon, *supra* note 67, at 304.

§4.05 AIR CARRIER CIVIL LIABILITY FOR DAMAGE CAUSED BY GNSS?

[A] Overview of the Regime of Air Carrier Civil Liability

The regime of air carrier civil liability is a core element of private international air law.[365] It regulates the legal relationship between an air carrier and the party suffering damage caused by an accident which took place during the carriage by air. According to this regime, the air carrier may be held liable for the compensation of damage by its passengers or persons on the ground. Compared with the regime of ATC civil liability (*see* §4.04), the regime of air carrier civil liability seems much more successful on the unification of private rules on international air carriage through a series of conventions and protocols concluded from Warsaw to Montreal via Rome (*see* §4.05[C] and §4.05[D]).

Legal instruments for air carrier civil liability provide, in most cases,[366] the basis for compensation of damage directly against an airliner under private law, regardless of whether damage was caused by a malfunction of a technical system including GNSS, human negligence, or force majeure, provided that the damage was caused by an accident, and other laid down in the international conventions on air carrier civil liability.[367] Accordingly, the necessity of a further GNSS civil liability regime was questioned, as some authors hold that the damage is already covered by existing provisions regulating an air carrier civil liability, particularly considering the fact that the sector of civil aviation has been indicated as one of the major domains of GNSS application.[368]

This section is designed to check whether or not this regime is adequate to deal with the issue of civil liability for damage caused in an air accident due to the malfunction of GNSS. For this purpose, the criteria used to comment on the regime of ATC civil liability are addressed here again, namely, whether victims may be able to get adequate compensation, and whether civil liability is channelled to the party that actually caused that damage through a series of recourse actions (*see* §4.04[D][5]).

365. Peter P.C. Haanappel, *The Law and Policy of Air Space and Outer Space: A Comparative Approach* (Kluwer Law International, 2003), at 67.
366. The exception here refers that the Warsaw Convention does not support a claim compensation against an air carrier, to which no fault could be obligated, in the case where damage was suffered by victims in an air accident caused by the failure of GNSS. *See also* §4.05[B] for further reasoning.
367. *Cf.* Kotaite, *supra* note 91, at 204.
368. UNIDROIT, *Item No. 7 on the Agenda: Third-Party Liability for Global Navigation Satellite System (GNSS) Services* (UNIDROIT, 2012), C.D. (91) 6, March 2012, at 5; Smith, *supra* note 153, at 606 & 607.

[B] The Legal Connection Between the Regime of Air Carrier Civil Liability and GNSS

In an air accident, the air carrier is always forced to go up against the claimants regardless of the causes of that damage,[369] yet depending on the applicable regime, that air carrier may be protected by a limited civil liability regime.[370] As discussed above, GNSS may be a sole cause or a contributing factor of an air accident which causes damage to passengers or third parties. In this case, the claimants may also sue the GNSS provider or ATC service provider to increase the likelihood of them getting full compensation, even though the GNSS provider and/or ATC service provider may currently be protected by the principle of State Immunity (*see* §4.04[D][4] and §5.03[B]).[371]

Technically speaking, GNSS may serve the operation of an aircraft through either an autopilot system on-board (*see* §4.02[B]), or an ATC service as a navigation aid (*see* §4.04[B]).

In the former case, legal relationships are illustrated in Figure 4.3. Victims may choose to sue the GNSS provider directly through line B, or, claim for civil liability against the air carrier through line A0, in which case the air carrier has to channel this burden to GNSS provider by a recourse action through line A1.

Figure 4.3 The Connection Between the GNSS Provider and the Air Carrier/Victims (I)

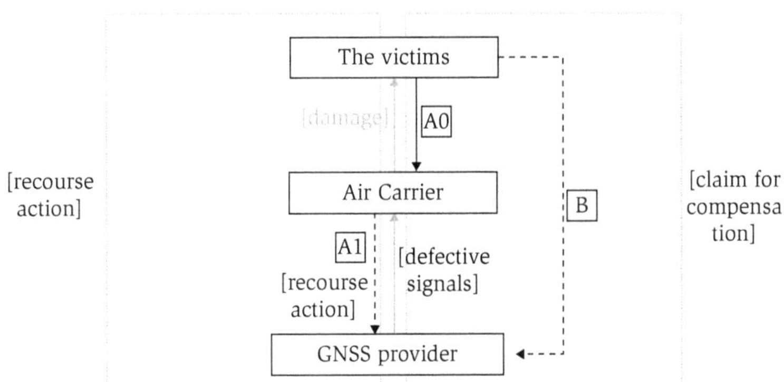

In the latter case, legal relationships are illustrated in Figure 4.4. Here, victims are facing three sets of legal proceedings:

(i) victims may claim compensation directly against the GNSS provider through line C.

369. von der Dunk, *supra* note 10, at 144.
370. *Cf.* Lagarrigue, *supra* note 253, at 10; Dempsey, *supra* note 78, at 238.
371. Schubert, *supra* note 301, at 258.

(ii) victims may sue the ATC service provider for the failure of ATC service through line B0, and then a recourse action through line B1 may make up for the interests of the ATC service provider who paid compensation. Since this process actually constitutes 'ATC civil liability' discussed in §4.04, the author will not repeat the discussion in this section.

(iii) victims may sue the air carrier following which the air carrier may engage recourse actions either to the GNSS provider directly through line A1, or to the ATC service provider through line A2 then finally to the GNSS provider through line A3.

Figure 4.4 The Connection Between the GNSS Provider and the Air Carrier/Victims (II)

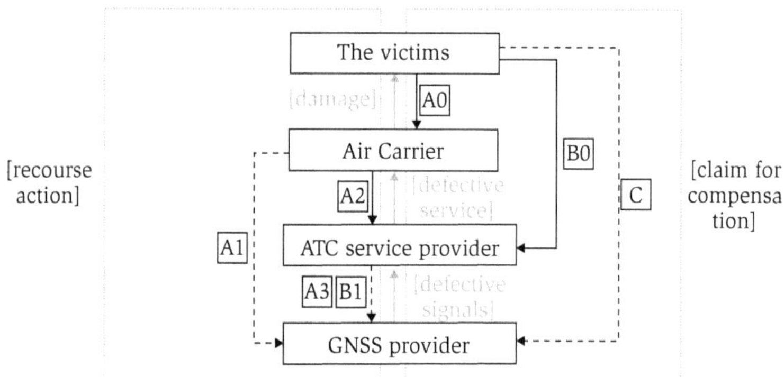

In brief, to get compensation for damage caused by an air accident due to defective GNSS signals, victims may choose (i) to claim directly against the actual wrongdoer, namely, the GNSS provider, or (ii) to sue the air carrier following which a series of recourse actions go from this air carrier to other parties so as to finally channel civil liability to an appropriate place, namely, the GNSS provider.

The next section examines whether present legal instruments governing air carrier civil liability are able to deal with the complicated system of direct action or recourse actions among victims, air carrier, ATC service provider, and GNSS provider.

[C] Contractual Liability: From Warsaw to Montreal

The contractual liability for international carriage by air was initially formed by the Convention for the Unification of Certain Rules Relating to Carriage by Air of 1929 (Warsaw Convention), and a complex system of amendments and additions[372] which all together are termed the 'Warsaw System'.[373] The Warsaw System can be basically

372. *See* Lawrence B. Goldhirsch, *The Warsaw Convention Annotated: A Legal Handbook* (Kluwer Law International, 2000), at 6-10.
373. Mendes de Leon, *supra* note 67, at 149-152.

concluded as a presumed fault-based civil liability regime encompassing breakable limits.[374] Under this regime, if claimants suffering damage caused by GNSS decide to sue the air carrier, they are exposed to the risk of inadequate compensation or no compensation at all, provided that the air carrier can demonstrate that it took all necessary measures to avoid the occurrence of damage, or that it was impossible to take such measures (*see also* §4.04[D][5][a]).[375] There is no case law demonstrating the feasibility of such actions.

Due to the rapid development of civil aviation, the Warsaw System was widely criticised for functioning unsatisfactorily,[376] and the Convention for the Unification of Certain Rules for International Carriage by Air of 1999 (Montreal Convention) was thus concluded as a modernised solution.[377] The Montreal Convention presents a great innovation by introducing a two-tier civil liability regime, that is (i) no-fault liability for damages not exceeding a certain amount of Special Drawing Rights (SDRs); and (ii) fault-based liability for damages exceeding that number of SDRs.[378]

Simply speaking, civil liability is limited where the fault is presumed, but it is unlimited where the fault is proven.[379] This limit of civil liability is currently set at 113,110 SDRs for death or injury of passengers.[380] Victims, namely passengers, never have to prove fault of any kind to recover the damages, sustained as a result of the bodily injury, of up to 113,110 SDRs; however, the air carrier may choose to rely on a defence for damages exceeding 113,100 SDRs.[381] This defence can be achieved by proving that "such damage was not due to the negligence or other wrongful act or omission of the carrier or its servants or agents;" or that "such damage was solely due to the negligence or other wrongful act or omission of a third party."[382] The burden of proof has been placed on the carrier.

The author argues that the two-tier civil liability regime only guarantees limited compensation which may merely cover a part, and not all of the damages in an air accident caused by the failure of GNSS, with the reasoning that:

(i) in the first tier, victims who suffered damage caused by GNSS are entitled to claim for compensation directly against the air carrier, but that compensation

374. *Ibid.*, at 151; Elmar Giemulla & Ronald Schmid (Eds), *Warsaw Convention: Commentary* (Kluwer Law International, 2005), at Chapter III WC – 1; Schubert, *supra* note 301, at 606.
375. Article 20 of the Warsaw Convention.
376. Malcolm Clarke, *Contracts of Carriage by Air* (Lloyd's List, 2010), at 16-17; Schubert, *supra* note 301, at 240; Tare Brisibe, *International Law and Regulation of Aeronautical Public Correspondence by Satellite* (Thesis of Leiden University, 2006), at 173.
377. Tompkins, *supra* note 79, at 33.
378. The limits of liability are counted by SDRs as defined by the International Monetary Fund, and those limits have to be reviewed and regularly revised in the context of the Montreal Convention. *See* Articles 23 & 24 of the Montreal Convention.
379. Elmar Giemulla & Ronald Schmid (Eds), *Montreal Convention: Commentary* (Kluwer Law International, 2010), at Article 22-5.
380. The limit of civil liability has been increased from 100,000 SDRs to 113,100 SDRs since 30 December 2009, which was first reviewed to remain the same in 2015. *See* ICAO, Working Paper C-WP/13478, the 188th session of ICAO Council, 7/10/09; ICAO, Electronic Bulletin EB 2014/035, 15 July 2014.
381. Mendes de Leon, *supra* note 67, at 181.
382. Article 21 of the Montreal Convention.

is limited to 113,100 SDRs. In other words, if the damage caused by GNSS is above 113,100 SDRs, the exceeding part is not protected by the regime of absolute liability in the first tier, and claimants have to go for the second tier (*see* below). For the part not exceeding 113,100 SDRs, the air carrier is entitled to any right of recourse against a third party - which here refers to the GNSS provider - for recovering the sums paid in compensation.[383] While Article 37 of the Montreal Convention makes it clear that any provisions of the Montreal Convention do not bar this right of recourse,[384] it remains silent in respect of the legal basis of any such recourse.[385] Similar with the discussion in §4.04[D][5], a legal link between the air carrier/ATC service provider and the GNSS provider is now missing to make that recourse action feasible, which needs to be established by either legal instruments regulating the relationship between the recourse creditor and the recourse debtor, or a contract containing the consensus on such a relationship (*see* §5.03[C]).[386]

(ii) in the second tier, the carrier may be fully or partly exonerated from civil liability to the extent of the damages exceeding 113,110 SDRs if the carrier proves that the damages are solely caused or contributed to by the GNSS providers. This will be very difficult to prove, but not impossible with the help of technical experts in the development of technology.[387] In this case, if the claim is *merely* based on the Montreal Convention, claimants cannot get full amount of compensation for the part exceeding 113,110 SDRs.

In brief, neither the Warsaw System nor the Montreal Convention is an adequate solution for civil liability in an air accident caused by defective GNSS signals, particularly in the case where damage exceeds 113,100 SDRs. Furthermore, the exclusivity of the Warsaw System and Montreal Convention, endorsed by series of court cases,[388] precludes the possibility of applying domestic laws on the rules of air carrier civil liability - in particular the issue of jurisdiction - when victims who suffered damage caused by GNSS claim for compensation against the air carrier.[389] As to the right of recourse entitled to the air carrier against a third party, that is a GNSS provider due to whose fault damage was caused, the Montreal Convention remains silent in

383. Article 37 of the Montreal Convention; UNIDROIT, *supra* note 46, at 24; UNIDROIT, *supra* note 368, at 6.
384. Article 37 of the Montreal Convention regulates 'Right of Recourse against Third Parties' reads as follows:

> Nothing in this Convention shall prejudice the question whether a person liable for damage in accordance with its provisions has a right of recourse against any other person.

385. Giemulla & Schmid, *supra* note 379, at Article 37-1.
386. *Cf.* Giemulla & Schmid, *supra* note 379, at Article 37-1.
387. UNIDROIT, *supra* note 46, at 24.
388. Pablo Mendes de Leon, *Jurisdiction under and Exclusivity of Private International Air Law Agreements on Air Carrier Liability: The Case of Airbus Versus Armavia Airlines (2013)*, in Pablo Mendes de Leon (eds), From Lowlands to High Skies: A Multilevel Jurisdictional Approach Towards Air Law (Martinus Nijhoff Publishers, 2013), at 269.
389. Mendes de Leon, *supra* note 67, at 172.

respect of the legal basis or specific rules of any such recourse, although it does not bar that right.

[D] Third-Party Liability: From Rome to Montreal

In contrast to the regime of air carrier contractual liability, the effort put into the rules for damage caused to third parties on the surface cannot be regarded as particularly successful.[390] Beginning with the failure in practice in the 1930s,[391] the Convention on Damage Caused by Foreign Aircraft to Third Parties on the Surface of 1952 (Rome Convention), followed by an amending protocol in 1978 to raise the extremely low liability caps,[392] was ratified as the first big success in establishing a uniform set of rules of third-party liability of air carriers.[393] However, compared with the Warsaw Convention and the Montreal Convention, the Rome Convention seems much less attractive for States and remains widely unadopted.[394] With the efforts led by ICAO, two new surface liability conventions – 'Unlawful Interference Compensation Convention' and 'General Risks Convention' – were introduced in 2009,[395] but neither convention is yet in force, nor has either convention received signatures or ratification from any significant air transport market.[396] Therefore, this section will only address the provisions of the Rome Convention.

For the issue of civil liability for damage caused in an air accident due to the failure of GNSS, most of those ratified and unratified conventions on air carrier third-party liability share more or less the same features or characteristics of the conventions under the regime of air carrier contractual liability.

First, the third-party liability conventions adopt a no-fault system,[397] and this is similar to the first tier of air carrier contractual liability under the Montreal Convention.

390. *Ibid.*, at 383-386.
391. An International Convention for the Unification of Certain Rules Relating to Damage Caused by Aircraft to Third Parties on the Surface was opened for signing in 1933 and entered into force in 1942, but only five States ever ratified it; A Protocol Supplementary to the Convention for the Unification of Certain Rules Relating to Damage Caused by Foreign Aircraft to Third Parties on the Surface was opened for signing in1938, but only received two parties. *See* Havel & Sanchez, *supra* note 74, at 315; Mendes de Leon, *supra* note 67, at 383.
392. Protocol to Amend the Convention on Damage Caused by Foreign Aircraft to Third Parties on the Surface Signed at Rome on 7 October 1952, entered into force on 25 July 2002, ICAO Doc 9257, 25/07/02.
393. Havel & Sanchez, *supra* note 74, at 315.
394. The numbers of parties of the Warsaw Convention, the Montreal Convention and the Rome Convention are 152, 131, and 50 respectively. *See* ICAO, *Current Lists of Parties to Multilateral Air Law Treaties*, https://www.icao.int/secretariat/legal/Lists/Current%20lists%20of%20parties/AllItems.aspx, last accessed 6 April 2018.
395. Convention on Compensation for Damage to Third Parties, Resulting from Acts of Unlawful Interference Involving Aircraft, 2 May 2009 done at Montreal, ICAO Doc 9920; Convention on Compensation for Damage Caused by Aircraft to Third Parties, 2 May 2009 done at Montreal, ICAO Doc 9919.
396. ICAO, *supra* note 394; Havel & Sanchez, *supra* note 74, at 320.
397. Article 1 of the Rome Convention.
An exception for this strict liability exists under Article 6 of the Rome Convention, but that is irrelevant which regard to negligence or other wrongful act or omission of the GNSS provider and is directed at the person who suffers damage.

In this case, a series of recourse actions is needed to channel civil liability from the air carrier to the GNSS provider, possibly via the ATC service provider.[398] Again, however, neither a legal link between the air carrier/ATC service provider and the GNSS provider, nor any specific *rules* for that recourse action is addressed in those third-party liability conventions, even though the *right* of recourse may be not denied there.

Second, the third-party liability conventions set a series of liability caps based on the weight of the aircraft,[399] and this is similar to the limits stipulated by the Warsaw Convention and the first tier of air carrier contractual liability under the Montreal Convention.[400] In this case, victims are again at the risk of inadequate compensation for the part exceeding those liability caps.

Third, although the third-party liability conventions provide the legal basis for claims based on a tort relationship or under an act in the absence of any contract,[401] this relationship is most of the time limited between an air carrier and third parties suffering damage, not being the term 'third party' used in Article 21 of the Montreal Convention where the GNSS provider may be involved as a liable party (*see* §4.05[C]). Therefore, the claim raised by victims directly against a GNSS provider is, in most cases, beyond the scope of application of the conventions governing the regime of air carrier third-party liability, since a legal link between victims and the GNSS provider is, again, missing under that system.

However, the focus here lies in Article 4 of the Rome Convention,[402] which may be called as an exception to the above paragraph. Article 4 of the Rome Convention obligates an unlawful user of an aircraft to be jointly and severally liable with 'the person entitled to its navigational control'. When GNSS is used as a navigation aid in the provision of ATC service, the ATC service provider may be viewed as 'the person entitled to its navigational control', but the GNSS provider is not an unlawful user in this regard as it is with the consent of the ATC service provider for the use of GNSS. In the case where GNSS signals are input into an autopilot system such as the GLS, a terrorist for example may be termed an unlawful user, but it is arguable whether a GNSS provider can be identified as 'the person entitled to its navigational control'. The

---

398. Article 10 of the Rome Convention provides that:

> Nothing in this Convention shall prejudice the question whether a person liable for damage in accordance with its provisions has a right of recourse against another person.

399. Article 11 of the Rome Convention.
400. Article 12 of the Rome Convention does allow those limits to be broken, however, only in instances where the damage was caused by a deliberate act or omission by the aircraft's operator or with an intent to cause damage; if the damage is solely caused by GNSS, those limits remain applicable.
401. Brisibe, *supra* note 376, at 182.
402. Article 4 of the Rome Convention reads as follows:

> If a person makes use of an aircraft without the consent of the person entitled to its navigational control, the latter, unless he proves that he has exercised due care to prevent such use, shall be jointly and severally liable with the unlawful user for damage giving a right to compensation under Article 1, each of them being bound under the provisions and within the limits of liability of this Convention.

author insists that the answer to that question depends on what kinds of signals are used in that GLS. For authorised GNSS signals, there will be a contractual relationship, either written or oral, between the air carrier and the GNSS provider, and the GNSS provider is reasonably entitled to the air carrier's – more specifically, the aircraft's – navigation control. For open GNSS signals, the air carrier uses GNSS signals in its GLS without the control of the GNSS provider. Since the GNSS provider actually has no way of rejecting being involved in the aircraft's navigational control, it is unfair in this case to obligate a GNSS provider of open signals to be liable for unlawful users such as terrorists.

In brief, most of those conclusions made under the regime of second-party liability of the air carrier about victims being inadequately ensured of getting compensation, and about the missing links between the air carrier, the ATC service provider and the GNSS provider similarly apply to the regime of air carrier third-party civil liability. Nevertheless, if an unlawful user of an aircraft hacks into the provision of authorised GNSS signals used in autopilot systems on-board, victims are entitled to hold the GNSS provider jointly or separately liable along with the unlawful user in accordance with Article 4 of the Rome Convention.

[E] Brief Conclusion

The regime of air carrier civil liability is indeed a selection and combination of presumed fault-based or no-fault, limited or unlimited, and, contractual or non-contractual (third party). In an air accident caused by the failure of GNSS, victims may seek compensation either by directly bringing a claim against the GNSS provider, or by suing the air carrier.

Most international legal instruments which were concluded from Warsaw to Montreal via Rome focus on engaging the liability of the air carrier, and they do not provide a legal basis for a direct claim raised by victims against the GNSS provider. However, if the claim is addressed against the air carrier pursuant to the terms of the applicable international convention, it may attempt to engage into recourse actions channelling the burden of civil liability to the GNSS provider. There is no case law giving evidence of the feasibility of such recourse actions. Furthermore, international air law usually restricts the extent of air carrier civil liability, which may not ensure victims receiving full compensation for damage caused by defective GNSS signals.

§4.06 CONCLUDING REMARKS

Civil liability is always one of the central issues in the legal theory and practice of civil aviation,[403] and it is not surprising that the international civil aviation community has contributed a great deal of effort into searching for solutions for GNSS civil liability during the implementation of CNS/ATM systems. The ICAO regime and Chicago

403. Milde, *supra* note 128, at 210.

Convention is for the public governance or technical assurance of GNSS implementation in the sense of public 'law', which is not appropriate for the issue of GNSS civil liability under private law.

When dealing with a case where an air accident was caused by the provision of defective GNSS signals, the regimes of ATC civil liability and of air carrier civil liability share the same pros and cons. Neither regime ensures adequate compensation for victims, particularly under the fault-based system of civil liability with caps. Nor does any of the regimes offer adequate legal links which may serve as legal basis for either a direct claim raised by victims against the GNSS provider, or a chain of recourse actions going from the air carrier to the ATC service provider and finally to the GNSS provider.

Put succinctly, international air law presents limited avenues for solving GNSS civil liability. Continuing efforts, coordination, and international cooperation are essential to finally paving a way forward on the issue of GNSS civil liability.

ANNEX The Technical, Institutional and Legal Evolution of CNS/ATM Systems under the ICAO Regime

Year	Events
1983	The ICAO Council established the FANS Committee.
1988	1. The FANS Committee developed the concept of CNS/ATM systems. 2. The ICAO Legal Committee started to work on the legal aspects of CNS/ATM systems, with a focus on GNSS. 3. The priority of legal aspects of CNS/ATM systems became Item 4.
1989	The FANS Phase II was established.
1991	1. The concept of CNS/ATM systems gained universal approval at the 10th Air Navigation Conference. 2. The 10th Air Navigation Conference requested the initiation of an agreement between the ICAO and GNSS-provider States concerning quality and duration of GNSS.
1992	1. The concept of CNS/ATM systems was endorsed at the 29th Session of the ICAO Assembly. 2. The priority of legal aspects of CNS/ATM systems moved to Item 5 and further to Item 1. 3. The 28th Session of the ICAO Legal Committee made preliminary conclusions on no inconsistency between the Chicago Convention and the implementation of the concept of CNS/ATM systems.
1993	The ICAO Air Navigation Commission established the Global Navigation Satellite System Panel (GNSSP, subsequently renamed NSP) to amend certain ICAO SARPs.
1994	1. The ICAO Council released the 'Statement of Policy on CNS/ATM Systems implementation and operation' for the implementation of CNS/ATM systems including GNSS. 2. The 29th Session of the ICAO Legal Committee: (1) prepared the Draft Agreement Between the International Civil Aviation Organization (ICAO) and GNSS Signal Provider Regarding the Provision of Signals for GNSS Services; (2) recommended establishing the LTEP, using a two-stage approach, namely, identifying a suitable solution for the immediate future, and a legal framework for the long-term future. 3. The US government and ICAO exchanged letters on the use of GPS in civil aviation.
1995	1. The 31st ICAO Assembly adopted Resolution A31-7 which requests the Council to establish the LTEP. 2. The LTEP was established by the ICAO Council.
1996	The Russian Federation and ICAO exchanged letters on the use of GLONASS in civil aviation.

Year	Events
1998	1. The first edition of the Global Air Navigation Plan for CNS/ATM Systems was released; 2. The World-wide CNS/ATM Systems Implementation Conference ((Rio de Janeiro) gave recommendations to legal action for CNS/ATM systems. 3. The 32nd ICAO Assembly: (1) adopted Resolution A32-19 'Charter on the Rights and Obligations of States Relating to GNSS Service', which was followed by a number of Recommendations offered by the LTEP on those subjects which need to be further studied before a consensus was reached; (2) adopted Resolution A32-20 'Development and elaboration of an appropriate long-term legal framework to govern the implementation of GNSS', which instructed the ICAO Council to establish a Secretariat Study Group on Legal Aspects of CNS/ATM Systems. 4. The ICAO Council established the Secretariat Study Group 'Development and Elaboration of an appropriate long- term legal framework to govern the implementation of GNSS'.
2001	The first package of SARPs was introduced in Volume I (Radio Navigation Aids) of Annex 10 (Aeronautical Telecommunications) to the Chicago Convention.
2002	The second edition of the Global Air Navigation Plan for CNS/ATM Systems was released.
2003	The 11th Air Navigation Conference recommended a worldwide transition to CNS/ATM systems.
2004	The Secretariat Study Group submitted its report, and the Group received approval on its accomplishing its mission at the 35th ICAO Assembly.
2005	1. The priority of legal aspects of CNS/ATM systems moved to item3. 2. The first edition of the GNSS Manual was released.
2007	1. The third edition of the Global Air Navigation Plan was released. 2. The US government and ICAO updated their exchanges of letters on the use of GPS in civil aviation.
2012	The 12th Air Navigation Conference addressed issues of use of multiple constellations and GNSS vulnerabilities.
2013	1. The fourth edition of the Global Air Navigation Plan was released. 2. The second edition of the GNSS Manual was released. 3. The priority of legal aspects of CNS/ATM systems moved to Item 3.
2014	The priority of legal aspects of CNS/ATM systems moved to Item 5.
2015	The priority of legal aspects of CNS/ATM systems moved to Item 4.
2016	The fifth edition of the Global Air Navigation Plan was released.
2017	The third edition of the GNSS Manual was released.
2018	The 13th Air Navigation Conference will pave the way forward to a more cost-efficient manner on the use of GNSS in civil aviation.

CHAPTER 5
The Way Forward for GNSS Civil Liability: Focusing on the Provision of Safety-of-Life Signals

§5.01 INTRODUCTION

History shows a deadlock in the search for solutions for GNSS civil liability. This notwithstanding, the application of GNSS in safety-critical sectors is increasing, which in turn leads to the growing risk of damage and concerns from the users' side. No entity can bear the failure of GNSS without sufficient remedy. It is irresponsible to wait for an accident to happen merely to justify the need for an appropriate GNSS civil liability regime. Instead, establishing a GNSS civil liability regime will make providers aware of their responsibility and liability, and urge them to better ensure the safety of applications. Moreover, GNSS providers themselves need legal certainty regarding the risk of civil liability in the long-term development of GNSS. Therefore, an appropriate approach is urgently required to pave the way forward for GNSS civil liability.

This chapter aims to achieve a highly practical solution for GNSS civil liability through a series of proposals where the rebalancing of interests between shareholders in the value chain of GNSS is always addressed. With the need for a fairness test on GNSS civil liability (see §5.02[A]), this chapter examines whether a free-of-charge policy (*see* §5.02[B]), an uncontrolled range of users for open signals (*see* §5.02[C]), and a disclaimer of civil liability (*see* §5.02[D]) justify GNSS providers not assuming civil liability (*see* §5.02). This chapter moves further to propose a roadmap (*see* §5.03) to satisfy the need for an international solution for GNSS civil liability (*see* §5.03[A]). For this purpose, (i) a legal and an institutional solution are presented to remove the obstacles placed by the doctrine of sovereignty immunity in realising GNSS civil liability in practice (*see* §5.03[B]); (ii) various solutions for GNSS civil liability are discussed to see which is more practical as a potential solution (*see* §5.03[C]); (iii) several international organisations are recommended for taking further action from different perspectives (*see* §5.03[D]); and (iv) concluding remarks round off this discussion (*see* §5.04).

§5.02 THE FAIRNESS OF GNSS CIVIL LIABILITY

[A] The Need for a Fairness Test on GNSS Civil Liability

Lives and properties could be endangered, thus the issue of civil liability will arise due to degraded GNSS signals below specific performance parameters (*see* §1.02[D]), as well as if no alert notice thereof is given within a specified timeframe.[1] Most GNSS providers however intend to question any reference to civil liability with the reasoning that GNSS signals are provided as public goods, free of charge, to an undetermined scope of potential users over which GNSS providers have no control, and with whom they are in no legal relationship apart from their duty to provide. GNSS provider States believe that open signals are provided 'as is',[2] and disclaimers of civil liability are always made.[3]

Therefore, this section examines the fairness of GNSS civil liability through three elements in the context of GNSS, namely: (i) free-of-charge policy, (ii) uncontrolled user range of open signals, and (iii) legal effect of a disclaimer of civil liability.

[B] Free-of-Charge Policy in the Context of GNSS

[1] Does a Free-of-Charge Policy Release the Civil Liability of GNSS Providers?

The public is widely impressed that GNSS signals of both core systems and augmented systems are available free of charge, which is in striking contrast with the financial pressure for the extremely high costs of GNSS development and operation. Against such a background, most GNSS providers show no additional courage to bear international civil liability for issues relating to faulty signals,[4] although they are confident that

1. European GNSS Agency (GSA), *ABOUT SoL*, https://egnos-user-support.essp-sas.eu/new_egnos_ops/services/about-sol, last accessed 10 July 2018.
2. 'As is' is a legal term and concept used to indicate that the goods are sold in their existing condition so as to disclaim and relieve the seller from liability for defects in that condition. Bryan A. Garner (Eds), *Black's Law Dictionary* (WEST, 2009), at 129.
3. Michael Milde, *Solutions in Search of a Problem? Legal Problems of the GNSS*, XXII(II) Annals of Air and Space Law 1997, at 211.
4. *Ibid.*; Paul B. Larsen, *Regulation of Global Navigation and Positioning Services in the United States*, in Ram S. Jakhu (Ed.), National Regulation of Space Activities (Springer, 2010), at 463; Andrea J. Harrington, *Regulation of Navigation Satellites in the United States*, in Ram S. Jakhu & Paul Stephen Dempsey (eds), Routledge Handbook of Space Law (Routledge, 2017), at 292; Ranjana Kaul, *Liability Implications of the Use of Global Navigation Satellite Systems (GNSS) for Communication, Navigation, Surveillance/Air Traffic Management (GNS/ATM) in Civil Aviation: With Special Focus on India*, XXXV(I) Annals of Air and Space Law 2000, at 431; Frans von der Dunk, *Navigating Safely Through the 21st Century: ICAO and the Use of GNSS in Civil Aviation*, 47 India Journal of International Law 2007, at 24.

the high quality of their systems constitutes a 'No Breakdown Guarantee' for all users.[5] Nevertheless, the author argues that such a free-of-charge policy in the context of GNSS does not justify a waiver of GNSS civil liability.

In practice, not all types of GNSS signals are produced against payment.[6] The free-of-charge policy of GNSS is mainly for open signals which, due to their technically unencrypted nature, makes it technically impossible to enforce payment on the provider's initiative if users do not cooperate.[7] It is only for those authorised signals which can be encrypted to control access that GNSS providers are able to charge fees, of which Galileo Commercial Service is a typical example.[8] Furthermore, the free-of-charge policy of GNSS is currently limited to the expression of 'free of direct user fees',[9] and whether a GNSS provider may charge fees through value-added service providers such as air navigation services providers (ANSPs) remains open (*see* below).[10] In addition, even the commitment to 'free of direct user fees' is made subject to the availability of funds.[11]

Even when taking a step back to say that all types of GNSS signals are available free of both direct and indirect user fees, a lower standard of civil liability should not be adopted simply because there is a lack of user charges.[12] To establish GNSS civil liability, what claimants need to prove is the existence of four required elements: the parties, unreasonable acts, damage, and causal link. Certain circumstances do exist to waive or mitigate civil liability (e.g., state of the art technology and acts of God), but none of them involves the question of whether or not GNSS signals are provided free of charge as a factor for consideration (*see* §2.04). GNSS providers should not be exempt from any civil liabilities caused by defective signals only because they provide signals for free.[13]

In private law, civil liability refers to the legal obligation of paying compensation for damage caused by a breach of duty either regulated by law, or agreed in a contract. The argument that a service or a signal is being offered free of charge does not justify

5. Battama Kantasuk, *General Legal Issues Concerning GNSS and the Impact on Developing Countries* (McGill LLM Thesis, 1997), at 56.
6. EU, Commission Implementing Decision (EU) 2017/224 of 8 February 2017, C/2017/0598; Frans von der Dunk, *Liability for Global Navigation Satellite Services: A Comparative Analysis of GPS and Galileo*, 30 Journal of Space Law (2004), at 132.
7. Scott Pace & Gerald Frost, et al., *The Global Positioning System: Assessing National Policies* (Rand, 1995), at 201.
8. EU, *supra* note 6; European GNSS Agency (GSA), *Galileo Services*, https://www.gsa.europa.eu/galileo/services, last accessed 21 May 2018.
9. ICAO, *Global Navigation Satellite System (GNSS) Manual*, Second Edition-2013, Doc 9849, AN/457, at 3-1, 3-2, and 4-5.
10. The term 'free of direct user fees' is apparently different from the normal expression of 'free of charge', and this gives rise to a series of questions, for example: (i) who is a direct user or an indirect user of GNSS? (ii) are indirect users provided GNSS signals free of charge? All relevant questions merit further research and discussion.
11. ICAO, *Report on the Establishing of a Legal Framework with Regard to CNS/ATM Systems Including GNSS*, A35-WP/75, LE/5, 28/07/04, at Attachment B; FAA, 2007 GPS and WAAS Service Commitments to ICAO, https://www.gps.gov/policy/cooperation/icao/2007-service-commitments.pdf, last accessed 21 May 2018.
12. Kantasuk, *supra* note 5, at 40.
13. *Ibid.*

the breach of a legal or a contractual obligation under most circumstances.[14] For example, if a car accident were caused by unreasonably poor conditions of a public road due to the negligence of the entity – which in most cases is the local government – responsible for maintaining the said road, this entity will be liable for the damage thereof.[15] The government has a general duty of care to keep the road reasonably safe, even though that road is available as a part of public infrastructure free of charge.[16] Similarly, the free-of-charge policy does not release GNSS providers from civil liability.

[2] Proposal for a Policy on Charges for GNSS Safety-of-Life Signals

Generally speaking, GNSS providers are under a duty of care to keep their signals either free or charged, reasonably safe for all users. The key here is what is considered as 'reasonable'. From a legal perspective, the charge policy is actually one of the influencing factors determining the level of duty of care of GNSS providers, just as the common understanding is that a toll highway provider is under a higher level of duty of care compared with a public road which is free of charge, and a commercial meteorological service should be more accurate than that provided by public agencies.[17] Therefore, the author proposes that a policy on charges or a cost-sharing mechanism is desirable for PNT signals used for safety-critical applications such as transportation, so that GNSS providers are obligated under a higher duty of care for safety-of-life signals.

The first focus being on GNSS safety-of-life signals,[18] the international civil aviation community has never shown a reluctance to share the costs of GNSS, even though, for the time being, GNSS providers do not request any financial commitments by non-provider States in the aviation sector.[19] Since the very beginning of the introduction of GNSS, ICAO has recognised that the issue of user charges will sooner or later arise in the long run.[20] Therefore, ICAO Resolution A32-19 'Charter on the Rights and Obligations of States Relating to GNSS Service' set out a basic principle of charge

14. Pablo Rodriguez-Contreras Perez, *GNSS Liability Issues: Possible Solutions to a Global System* (McGill University, 2002), at 58; UNIDROIT, *Item No. 7 on the Agenda: Third-Party Liability for Global Navigation Satellite System (GNSS) Services* (UNIDROIT, 2012), C.D. (91) 6, March 2012, at 6.
15. Larry W. Thomas, *Liability of State Highway Departments for Design, Construction, and Maintenance Defects*, 4 Selected Studies in Highway Law 1978, at 1771.
16. *See also Basia Lejonvarn v. (1) Peter Burgess (2) Lynn Burgess* [2017] EWCA Civ 254, in which case: An architect who provided professional services to her friends, free of charge and without a contract, nevertheless owed a duty of care to exercise reasonable care and skill and was therefore legally responsible and liable for her negligence.
17. A US court argued that weather forecast is a classic example of a prediction of indeterminate reliability rather than an accurate science so that the governmental agency providing inaccurate forecast should not be held liable. However, this case in turn demonstrates that GNSS as an accurate science should be legally responsible and liable for its inaccuracy. *See Brown v. United States*, 790 F.2d 199 (1st Cir.1986).
18. von der Dunk, *supra* note 6, at 150.
19. Milde, *supra* note 3, at 208.
20. Assad Kotaite, *ICAO's Role with Respect to the Institutional Arrangements and Legal Framework of Global Navigation Satellite System (GNSS) Planning and Implementation*, XXI(II) Annals of Air and Space Law 1996, at 203.

policy on GNSS for the use of civil aviation which requires that "any charges for GNSS services shall be made in accordance with the Chicago Convention."[21]

Facing the incremental costs of GNSS, in 2007 the ICAO Council disseminated provisional policy guidance by a State Letter to its Contracting States on GNSS cost allocation.[22] This provisional guidance declares, among other things, that basic GNSS services will be provided free of charge as a public good, while more advanced GNSS services (including augmentation services) requiring a higher quality of service, and hence higher costs, will in most cases have to be paid.[23] Furthermore, the costs for GNSS in the civil aviation section should be distributed among ANSPs and on the different phases of flights in accordance with existing ICAO policy and guidance, and ANSPs then recover the costs from the users within their existing charging systems.[24] Notably, ICAO recognised the link between the issue of GNSS civil liability and the mechanism of GNSS cost allocation by stating that:

> Once a consensus has been reached on the definition of basic services and liabilities of GNSS service providers, this provisional guidance is to be redrafted with appropriate wording for inclusion in *ICAO's Policies on Charges for Airports and Air Navigation Services* (Doc 9082).[25]

In brief, the international civil aviation community keeps an open mind on GNSS cost allocation and distribution. With a clear roadmap, ICAO is prepared to incorporate GNSS into the existing charge policy on ANSs immediately once the regime of GNSS civil liability is agreed. Furthermore, fees charged for the use of GNSS may in turn be used for the operation and maintenance of GNSS,[26] which is helpful in ensuring a higher degree of safety for GNSS applications in safety-critical sectors.

In addition, considering the huge economic interests enhanced by GNSS,[27] the author thus suggests that GNSS signals - in particular augmented signals - for safety-of-life applications may be subject to a policy on charges, while keeping the free-of-charge policy on open signals in other domains.[28] Payment may be collected either indirectly through value-added service providers such as ANSPs from end users

21. ICAO, *Assembly Resolutions in Force (as of 6 October 2016)*, Doc 10075, at V-10.
22. ICAO, *Infrastructure Management: GNSS – Cost Allocation*, https://www.icao.int/sustainability/Pages/eap-im-gnss-cost-allocation.aspx, last accessed 22 May 2018.
23. *Ibid.*
24. *Ibid.*
25. *Ibid.*
26. For example, in the case of GAGAN, the AAI collects users fees as both an ANSP and GAGAN signal provider according to the tariff regulated by Airports Economic Regulatory Authority of India, which is the economic regulator of the airport infrastructure sector and air navigation services in India; Meanwhile, GAGAN is a joint venture of the AAI and the ISRO which is the operator of GAGAN, in which case the user fees collected by the AAI may somehow reinvested or reused on the development, operation and maintenance of GAGAN. *See* Kaul, *supra* note 4, at 435-437; ISRO, Satellite Navigation, https://www.isro.gov.in/spacecraft/satellite-navigation, last accessed 22 May 2018.
27. Milde, *supra* note 3, at 196.
28. The author does not distinguish between core GNSS signals and augmented signals for a policy on charges, as the cost of both of them needs to be allocated and shared with all stakeholders, particularly considering the corporatisation and commercialisation of GNSS, for example, the case of Galileo.

such as airliners,[29] or directly from users if an agreement exists between providers and those users. This charge policy may eliminate the resistance, and increase the affordability of GNSS providers to civil liability. For example, it is expected that a substantial element of the proposed package of paid services in the context of Galileo would be the inclusion of civil liability acceptance on the part of the operator.[30]

If there is no way to share the cost of GNSS development and operation with users, providers might look for protection against civil liability for damage as a result of their element-input, rather than being required to offer protection to such elements regarding civil liability.[31]

Furthermore, the policy allowing GNSS providers to recover their costs may accelerate the corporatisation or commercialisation of GNSS where the doctrine of sovereign immunity does not apply in civil suits concerning GNSS civil liability (*see* §5.03[B][4]).

[C] The Problem of Open Signals in the Context of GNSS Civil Liability

[1] The Classification of GNSS Signals

GNSS signals could be technically categorised either as open signals, or as authorised signals according to whether those signals are openly accessible or not.[32] Most GNSS providers provide both open signals and authorised signals, but the former seemingly connect much more with our daily life.[33] Open signals are usually provided based on a free-of-charge policy, but this is not always true. Authorised signals can also be provided free of charge as long as providers have agreed. Moreover, open signals may also be charged user fees through a contract on the user's initiative for a service guarantee, despite the technically unencrypted nature of open signals.

GNSS open signals present additional legal challenges due to their open access to the public. Any user with a compatible receiver can use unencrypted open signals, but the GNSS provider has no information about, and control over the user group. This particularity of undetermined scope of potential users, as well as the free-of-charge policy (*see* §5.02[B]), makes GNSS providers reluctant to accept any responsibility for a warranty on open signals and civil liability for damage caused by defective open

29. von der Dunk, *supra* note 6, at 150.
30. Frans von der Dunk, *Space Law and GNSS – A Look at the Legal Frameworks for 'Outer Space'*, May/June InsideGNSS 2017, at 38.
31. *Cf.* von der Dunk, *supra* note 6, at 153.
32. In practice, the terms open signal and open service are usually used interchangeably, but they may also be different sometimes. For example, in the context of EGNOS, both Open Service and SoL Service are based on open signals which are openly accessible. GSA, *ABOUT OS*, https://egnos-user-support.essp-sas.eu/new_egnos_ops/services/about-os, last accessed 25 May 2018; GSA, *supra* note 1.
33. Except Galileo Commercial Service, signals provided for civilian uses by core systems such as GPS, GLONASS, and BDS and augmented systems such as EGNOS and GAGAN are with open access.

signals.[34] GNSS providers thus warn users that they use GNSS signals at their own risk through certain disclaimers of civil liability (*see* §5.02[D]).

In this regard, the author agrees that GNSS providers are justified in refusing contract liability of open signals, but also argues that GNSS providers must keep a minimum duty of care in tort law for the provision of open signals. For the maximum level of safety-critical applications, the author proposes an authorisation mechanism on GNSS safety-of-life signals for a quality guarantee.

[2] Is There a Contractual Relationship in the Provision of Open Signals?

The existence of a contract or contractual relationship determines the establishment of contractual liability. Although the author does not rule out the possibility of signing a contract on the user's initiative with GNSS providers for the provision of open signals, such as EGNOS Working Agreements (EWA) for the application of EGNOS SoL signals with open access,[35] the practice that both individual and professional users treat, as a given, the ubiquitous availability without discrimination seems far more common,[36] for which no written contract or agreement was concluded.

In this regard, scholars as well as GNSS providers hold that the provision of open signals does not evoke any contractual liability since it is not contracted for.[37] Other opinions however put forward the establishment of a virtual or quasi-contractual relationship,[38] or an implied contract because of expected use,[39] breach of which gives users a remedy based on contractual liability.[40] The author supports the former viewpoint that an implied contract does not arise in the provision of open signals.[41]

From a private law perspective, an offer by the offeror, and the acceptance of this offer by the offeree are two essential elements to establish a binding contract,[42] regardless of whether that contract constitutes a written, oral, or implied contract.[43] Most GNSS providers have issued public documents describing the access policy and

34. Milde, *supra* note 3.
35. GSA, *EGNOS Safety of Life (SoL) – Service Definition Document*, Revision 3.1, 26/9/2016, at 22.
36. Alessandra A.L. Andrade, *The Global Navigation Satellite System* (Ashgate, 2001), at 109; Ingo Baumann, *State of Play in the European Union: Liability for GNSS Signals and Services*, November/December InsideGNSS 2015, at 38.
37. Baumann, *ibid.*, at 39; von der Dunk, *supra* note 6, at 140, 153 & 156; Alessandro del Ninno, *Providing GNSS Services: The Legal Perspective. The Existing Regime and Its Shortcomings with Regards to Liability, Data Policy and Data Integrity*, presented to The International Workshop GNSS Technology Advances in a Multiconstellation Framework (Rome, 26 September 2014).
38. Alessandra Arrojado Lisboa de Andrade, *Navigating into the New Millennium: The Global Navigation Satellite System Regulatory Framework* (McGill LLM Thesis, 2000), at 88.
39. UNIDROIT, *supra* note 14.
40. Kotaite, *supra* note 20.
41. It is argued that an implied contract exists between users and such application provider as Google Maps or TomTom, as users have to agree to the terms and conditions presented in the application before they are allowed to use that App. (*See* UNIDROIT, *supra* note 14.) But that implied contract does not qualify as a contact discussed here which must be concluded between GNSS providers and users (*see* §2.02).
42. Jaap Hage, et al. (Eds), *Introduction to Law* (Springer International Publishing Switzerland, 2014), at 57.
43. The term implied contract is defined as:

performance parameters of their open signals, and some GNSS providers have exchanged letters with user communities such as ICAO for that purpose (*see* §4.03[C]). Do any of these actions constitute an offer in the legal sense? The answer is no. An offer can only exist when it reflects the intention to be legally bound and is sufficiently clear about the contents of the resulting contract.[44] As discussed, most of those public documents do not present GNSS providers' legal opinion and intention to conclude a contract, and the exchange of letters is merely directed towards a political commitment (*see* §4.03).

Even when taking a step back to say that the public statements made by GNSS providers on the availability of their open signals legally constitutes an offer, users – as the offeree – have to dispatch their acceptance to those GNSS providers to form a binding contract. Importantly, it is very relevant that parties know at what moment the acceptance reaches the offeror– GNSS providers,[45] but indeed in most cases this is unrealisable for open signals through technical means due to their unencrypted nature and passive-positioning model where open signals are transmitted one way from GNSS satellites to receivers.[46] These technical features of open signals make it impossible for GNSS providers to know and monitor who receives and uses or misuses those signals. From a legal perspective, it means that GNSS providers cannot know who is the other contracting party, and the acceptance is impossible to be dispatched from GNSS users to providers.[47]

Therefore, the provision of open signals would not indicate the existence of an implied contract.[48] GNSS providers neither own implied contractual obligations, nor bear contractual liability to users of open signals unless the contract was concluded specifically for the provision of open signals.

[3] *The Minimum Duty of Care in Tort Law for the Provision of Open Signals*

In a case where a sufficiently serious accident was caused by defective GNSS open signals with a sufficiently large number of claimants, efforts would have to be made to circumvent the issue of no one being liable because there was no contract.[49] Although the provision of GNSS open signals does not recognise the existence of an implied

---

> A contract not created by express words but inferred by the courts either from the conduct of the parties or from some special relationship existing between them.

Elizabeth A. Martin (Eds), *A Dictionary of Law* (Oxford University Press, 2003), at 243.
44. Hage, *supra* note 42, at 58.
45. *Ibid.*, at 59.
46. Elliott D. Kaplan & Christopher Hegarty (Eds), *Understanding GPS/GNSS: Principles and Applications* (Artech House, 2017), at 89.
47. Kantasuk, *supra* note 5, at 57.
48. Ninno, *supra* note 37.
49. UNIDROIT, *supra* note 14.

contract, it also does not deny in principle the possibility of a claim for non-contractual liability mainly under tort law (*see* §2.03[C]).[50]

In tort law, most questions are addressed under the heading of duties of care,[51] whose existence encourages a person to be careful.[52] A duty of care refers to the legal obligation, regulated by either judicial decisions, or statutes, to take care, in most situations where one can reasonably foresee that his or her actions may cause physical damage to the person or property of others.[53] Most legal systems recognise the remedies for civil wrong for the breach of a duty of care.[54] A duty of care does not disappear merely because of open access to services or products. For example, the caretaker of a roadway with open access to the public has to bear civil liability for damage caused by unreasonable negligence in the roadway's maintenance;[55] providers of open access Wi-Fi were obligated to be liable for unlawful behaviour of anonymous users in certain jurisdictions, despite the fact that opposing opinions exist.[56]

GNSS open signals have now penetrated every corner of daily life, thus a status of reliance on those signals has formed. In this case, users are reasonable to expect and trust that open signals are provided as always according to normal performance parameters released to the public, unless a warning or notification on an abnormal situation were made timely. If there is a special relationship and a reasonable reliance interest, a duty of care will arise.[57] Therefore, providers are under a duty of care to avoid causing damage in the provision of open signals,[58] and they ought to reasonably foresee that the defects or failure of those signals *without warning* may cause loss of life and damage to property, particularly in safety-critical domains.[59]

Nevertheless, the author argues that the duty of care imposed on providers of open signals is at a minimum level. In other words, the purpose to place a duty of care in tort law is to avoid causing damage to others, rather than to force or 'kidnap' GNSS providers into providing open signals at a safety level reaching the requirements of certain groups of users. GNSS providers are in no way always obligated to provide open signals as offered in the unilateral statements.[60]

50. Ninno, *supra* note 37; von der Dunk, *supra* note 6, at 153.
51. Hage, *supra* note 42, at 107.
52. Ronald B. Standler, *Tort Liability in the USA for Negligent Weather Forecasts*, http://www.rbs2.com/forecast.pdf, last accessed 21 May 2018.
53. Martin, *supra* note 43, at 165.
54. Muhamed Mustaque, *Legal Aspects Relating to Satellite Navigation in Air Traffic Management with Specific Reference to GAGAN in India*, 50 Proceedings on the Law of Outer Space 2008, at 340.
55. Thomas, *supra* note 15.
56. Christoph Busch, *Secondary Liability for Open Wireless Networks in Germany: Balancing Regulation and Innovation in the Digital Economy*, in Graeme B. Dinwoodie (eds), Secondary Liability of Internet Service Providers (Springer, 2017), at 364; Casey G. Watkins, *Wireless Liability: Liability Concerns for Operators of Unsecured Wireless Networks*, 65(2) Rutgers Law Review 2013, at 638.
57. *See Hedley Byrne & Co Ltd v. Heller & Partners Ltd* (1964) AC 465 (HL); Robert James Hollyman, *Liability and Reliability: The Reliance Interest in Negligence Damages* (University of Toronto LLM Thesis, 1999), at 9 & 23.
58. Nick Ward, *Monitoring the Integrity of GNSS*, 47 (2) Journal of Navigation 1994, at 185.
59. *Cf.* Ruwantissa Abeyratne, *Space Security Law* (Springer, 2011), at 25.
60. Kantasuk, *supra* note 5, at 57.

As discussed in §4.03, most service commitments made in national policy, or technical documents or letters exchanged with international communities such as ICAO are not legally binding. Therefore, unless a service guarantee is legally binding through the conclusion of a contract or written into a national or international legal instrument, providers are at liberty to temporarily or permanently degrade or terminate the provision of open signals as long as they perform the duty to warn or notify users of those decisions in a reasonable manner which gives a reasonable period of time for users to take measures to avoid causing damage. According to the principle of good faith, which is one of the general principles of law recognised by civilised nations,[61] what the duty of care regulates is the reliance and confidence that GNSS users may place on providers for the provision of open signals; once GNSS users have been reasonably warned or notified of the altered circumstances in the provision of open signals, they are no longer entitled to claim for any indemnification for further losses.[62]

It is argued that a positive correlation exists between the degree of reliance on GNSS and the expectation on the assurances for its accessibility and reliability.[63] For example, GNSS has been proposed as a sole means of navigation in the civil aviation sector, and this factor was called for the considerations in establishing the legal framework for GNSS, in particular the regime of civil liability.[64] From the perspective of users, the first condition for accepting that proposal is always confidence in the reliability of GNSS.[65] Due to the lack of such confidence, the civil liability problem has been particularly addressed to being solved before GNSS becomes the sole means of navigation.[66] This notwithstanding, the author argues that the proposal for GNSS as the sole means of navigation does increase the degree of reliance on GNSS, even though it is irrelevant to the level of duty of care imposed on GNSS providers of open signals.

In tort law, a person owes a duty of care to avoid causing damage due to his/her acts, but he/she does not have a general duty to benefit others if he/she did not agree

61. Steven Reinhold, *Good Faith in International Law*, 2 UCL Journal of Law and Jurisprudence 2013, at 41
62. *Cf.* Bin Cheng, *General Principles of Law as Applied by International Courts and Tribunals* (Cambridge University Press, 1994), at 137.

In the book, Professor Bin Cheng describes the principle of good faith by the following words:

> If State A has knowingly led State B to believe that it will pursue a certain policy, and State B acts upon this belief, as soon as State A decides to change its policy-although it is at perfect liberty to do so-it is under a duty to inform State B of this proposed change. Failure to do so, when it knows or should have known that State B would continue to act upon this belief, gives rise to a duty to indemnify B for any damage it may incur.

63. Chiara Lucchini Gilera, *GNSS Third-Party Liability: The European Experience of Galileo*, 49 Proceeding on the Law of Outer Space 2006, at 459; Ludwig Weber & Jiefang Huang, *ICAO and GNSS*, 3(1) Outer Space Committee Newsletter 2000, at 45; Ludwig Weber, *The Global Navigation and Communications Satellite Systems and the Role of ICAO*, in: ESA/ECSL, et al., Proceedings of the Third ECSL Colloquium – International Organisations and Space Law (European Space Agency, 1999), at 101.
64. Weber, *ibid.*
65. EUROCONTROL, *GNSS Sole Service Feasibility Study*, EEC Note No. 04/03, Project GNS-Z-SBAS, at 115.
66. Kantasuk, *supra* note 5, at 55.

to this.[67] In the absence of a legally binding service guarantee which may be achieved through an authorisation mechanism (*see* §5.02[C][4]), users are at their own risk in relying solely on unwarranted open signals.[68] Based on tort law (not contract law), users should be aware that a provider is free to degrade quality or close the access as long as this provider performs its duty of care to avoid causing damage by issuing a reasonable warning or notification for the abnormal situation in the provision of open signals.

Furthermore, it is argued that the public nature of GNSS open signals obligates providers under a duty of care of *particularly high standards* toward users.[69] The development and operation of GNSS require significant public resources and finance;[70] GNSS is in turn built as public infrastructure. Therefore, it seems to be general consensus that GNSS open signals qualify as global public goods (GPG) or services.[71] GPG are goods whose benefits and/or costs extend to all countries, people, and generations.[72] A question may arise here whether there is a link between the term 'GPG' and the issue of civil liability.[73] The author argues that the public nature of GNSS only confirms the *status* of reliance on open signals by the public, and the identity as a type of GPG only indicates that the scope of that reliance has been extended to a *global* scale. At least until the costs of GNSSs are shared globally, whether and how to provide open signals - although those signals qualify as GPG - is at the freedom of providers as long as they do not violate their duties of care.

In brief, a GNSS provider of open signals owes a general duty of care in tort law to GNSS users, such as airliners, but the *status* of reliance, rather than the *degree* of reliance is relevant for that duty of care. A GNSS provider for open signals is merely obligated under a duty of reasonable and timely notification, not keeping high-standard

67. Hollyman, *supra* note 57, at 15.
68. UNIDROIT, *An Instrument on Third Party Liability for Global Navigation Satellite System (GNSS) Services: A Preliminary Study* (UNIDROIT, 2010), Study LXXIX - Preliminary Study, March 2010, at 46.
69. Gilera, *supra* note 63.
70. Baumann, *supra* note 36.
71. Serge Plattard, *Can Global Navigation Satellite Systems Signals Qualify to Become a World Public Good?*, 3(3) New Space 2015, at 142; Rajeswari Pillai Rajagopalan & Narayan Prasad (Eds), *Space India 2.0: Commerce, Policy, Security and Governance Perspectives* (Observer Research Foundation, 2017), at 171; Michael Chatzipanagiotis & Konstantina Liperi, *Regulation of Global Navigation Satellite Systems*, in Ram S. Jakhu & Paul Stephen Dempsey (eds), Routledge Handbook of Space Law (Routledge, 2017), at 174; US National Oceanic and Atmospheric Administration (NOAA), *GPS: The Global Positioning System: A Global Public Service Brought to You by the US Government*, https://www.gps.gov, last accessed 30 May 2018.
72. Inge Kaul & Ronald U. Mendoza, *Advancing the Concept of Global Public Goods*, in Pedro Conceição, Katell Le Goulven & Ronald U. Mendoza (eds), Providing Global Public Goods: Managing Globalization (Oxford University Press, 2003), at 95.
73. What matters concerning the status of GPG matters seems to be the need for an international solution to GNSS civil liability (*see* §5.03), since GPG is defined as "issues that are broadly conceived as important to the international community, that for the most part cannot or will not be adequately addressed by individual countries acting alone and that are defined through a broad international consensus or a legitimate process of decision-making." The International Task Force on Global Public Goods, *Meeting Global Challenges: International Cooperation in the National Interest* (International Task Force on Global Public Goods, 2006), at 13.

safety requirements, for abnormal situations departing from its normal statement on performance parameters in the provision of open signals.

[4] *Proposal for an Authorisation Mechanism on GNSS Safety-of-Life Signals*

Unless a contract was expressly concluded (see below), no contractual liability may be attributed to GNSS providers in the provision of open signals, since an implied contract also does not exist (*see* §5.02[C][2]). GNSS providers of open signals do owe a duty of care in tort law to GNSS users, but it only refers to the obligation of notification, rather than maintaining a stable performance standard meeting the requirement of users. GNSS providers are free to make any decisions on the provision of open signals as long as they reasonably warn or notify GNSS users in advance of that change. This notwithstanding, the insufficiency of a civil liability regime and instability in the provision of open signals may be not very concerned by public applications, although they cannot satisfy safety-critical applications – for example, aviation navigation – as the impact of loss of navigation capability is not only on a single aircraft, but on a predetermined population of aircraft in a specified airspace.[74]

Therefore, a guarantee for indemnification and a 'genuine' duty of care, which is not limited to notification but constitutes a stable service guarantee with legally binding effect, is advisable to be established in the provision of safety-of-life signals. For that purpose, in the near future the author does not expect that GNSS providers will regulate service commitments in national legislation which may serve as the basis for tort liability, but proposes an authorisation mechanism which may be achieved through either a written contract for the provision of open signals, or an implied contract for the provision of authorised signals.

For the first approach, GNSS safety-critical applications continue to rely on open signals, but an express contract must be concluded between providers and users since an implied contractual relationship does not exist (*see* §5.02[C][2]). The performance parameters and the terms of civil liability in the provision of open signals would need to be agreed by the parties to that contract. In practice, the experience of EWA, where a service guarantee and the term contractual liability are addressed,[75] may be expanded to the world.

---

74. ICAO, *Performance-based Navigation (PBN) Manual*, Volume II. Implementing RNAV and RNP, Third Edition-2008, Doc 9613, AN/937, at II-A-3-2.
75. The EWA includes:

> EWA contractual document: The agreement itself containing contractual liability with two annexes:
>
> Annex 1: Including the "ESSP SAS SoL Service Commitment" as stated in this EGNOS SoL SDD. It also includes reference to contingency coordination between ESSP and the ANSP.
> Annex 2: Including the "Service Arrangements" defined between the ESSP and the ANSP with the purpose to enable the ANSP to implement Performance Based Navigation (PBN) procedures based on EGNOS [...].

GSA, *supra* note 35, at 23.

For the second approach, GNSS safety-critical applications would need to change to utilise authorised signals. Different from open signals, authorised signals deny unauthorised access for security reasons. At the time being, authorised signals are mostly restricted to military and commercial users.[76] Yet, civilian safety-of-life signals are usually not encrypted, which exposes them to harmful interferences and spoofing technologies[77] through which terrorists may target civil aircraft navigating on GNSS open signals.[78] Technically speaking, applying an authorisation mechanism for GNSS safety-of-life signals is feasible in practice due to the fact that certain GNSS providers leave certain frequencies of civil signals exclusively for the use of safety-critical sectors such as civil aviation,[79] meaning that the requirement of authorisation for GNSS safety-of-life signals does not deny access to open signals for the public applications. Legally speaking, providing and using authorised signals is a matter of an implied contract. Controlled or closed access to GNSS safety-of-life signals requires users and providers to communicate about the rights and obligations of each other, which may include a policy on charges and the terms of civil liability. Despite the parties not having concluded a written contract on the provision of GNSS safety-of-life signals, the application submitted by users and the approval given by providers for the encrypted code indeed qualify as an offer and acceptance which form an implied contract.

[D] Legal Effect of Disclaimer of Civil Liability

[1] Introduction to Disclaimer of Civil Liability

A disclaimer of civil liability,[80] variously known as an exemption or an exclusion clause,[81] is an oral or a written notice that intends to negate or limit the party

76. *See* Paul D. Groves, *Principles of GNSS, Inertial, and Multisensor Integrated Navigation Systems* (Artech House, 2013), at 312.
77. Xihui Chen, et al., *A Trust Framework for Evaluating GNSS Signal Integrity*, in Patrick Kellenberger (eds), Proceedings of 2013 IEEE 26th Computer Security Foundations Symposium (The Institute of Electrical and Electronics Engineers, Inc., 2013), at 179.
78. Yvon Henri, *Preventing Harmful Interference to Satellite Systems*, http://news.itu.int/preventing-harmful-interference-satellite-systems/, last accessed 31 May 2018.
    This factor was also viewed as an important one by the FAA on understanding the feasibility of GNSS as a sole means of navigation (*see* §5.02[C][3]). Francis P. Schubert, *An International Convention on GNSS Liability: When Does Desirable Become Necessary?*, XXIV Annals of Air and Space Law 1999, at 262.
79. *See* NOAA, *New Civil Signals: Third Civil Signal: L5*, https://www.gps.gov/systems/gps/modernization/civilsignals/, last accessed 31 May 2018; GSA, *Service Access: Safety of Life Service (SoL)*, https://www.gsa.europa.eu/european-gnss/egnos/services/service-access, last accessed 31 May 2018.
80. It is necessary to distinguish the term 'disclaimer of civil liability' and 'indemnification clause'. The latter refers to a remedy that allows one person to recover reimbursement from another upon the happening of an event. Paula Duggan Vraa & Steven M. Sitek, *Public Policy Considerations for Exculpatory and Indemnification Clauses: Yang v. Voyagaire Houseboats*, 32(4) William Mitchell Law Review 2006, at 1321.
81. Boakye Danquah Kofi Henaku, *The International Liability of the GNSS Space Segment Provider*, XXI(I) Annals of Air and Space Law 1996, at 155.

expressing the disclaimer from any civil liability with the particular damage described.[82] A disclaimer of civil liability is generally made in a bilateral agreement through one or more terms with precise and conspicuous language to be effective under contract law,[83] even though it may also arise by a unilateral statement on warnings or expectations to the general public under tort law, either common law or statute.

In practice, a disclaimer of civil liability may be briefly stated by the use of specific idioms such as the term 'as is' or 'with all faults'.[84] Disclaimers of civil liability are broadly used in the satellite telecommunications industry in the case of signal failure due to telecommunications breakdowns,[85] such as Article XII of the Operating Agreement on the International Maritime Satellite Organization,[86] and Article 36 of the Constitution of the International Telecommunication Union.[87]

[2] Disclaimers of Civil Liability in the Context of GNSS

In most cases, GNSS providers try to deny potential civil liability for the provision of open signals through a unilateral disclaimer[88] since those signals with free access are not contracted for (*see* §5.02[C][2]). Although the unilateral disclaimer can be oral or written, formal or informal, it is usually included in the technical documents describing performance standards and interface specifications of GNSS signals.[89] For example, Japan's Quasi-Zenith Satellite System (QZSS) presents a 'Disclaimer of Liability' at the very beginning of its interface specification document for both the use of the document, and satellite positioning services and message services provided by the QZSS;[90] both

82. Garner, *supra* note 2, at 531.
83. *Ibid.*; Don Tracy, *Disclaiming and Limiting Liability for Commercial Damages*, 83 Commercial Law Journal 1978, at 14.
84. Section 2-316(3)(a) of the US Uniform Commercial Code.
85. ICAO, *Progress Report on the Establishment of a Legal Framework with Regard to CNS/ATM Systems Including GNSS*, A33-WP/34, LE/5, 22/06/01, at 4.
86. Article XII of the Operating Agreement on the International Maritime Satellite Organization (INMARSAT) reads as follows:

> Neither the Organization, nor any Signatory in its capacity as such, nor any officer or employee of any of them, nor any member of the board of directors of any Signatory, nor any representative to any organ of the Organization acting in the performance of their functions, shall be liable to any Signatory or to the Organization for loss or damage sustained by reason of any unavailability, delay or faultiness of telecommunications services provided or to be provided pursuant to the Convention or this Agreement.

87. Article 36 of the Constitution of the International Telecommunication Union reads as follows:

> Member States accept no responsibility towards users of the international telecommunication services, particularly as regards claims for damages.

88. *Cf.* Henaku, *supra* note 81, at 154.
89. Souichirou Kozuka, *Regulation of Navigational Satellites in Japan*, in Ram S. Jakhu & Paul Stephen Dempsey (eds), *Routledge Handbook of Space Law* (Routledge, 2017), at 311.
90. *See* Cabinet Office of Japan, *Quasi-Zenith Satellite System – Interface Specification – Positioning Technology Verification Service*, IS-QZSS-TV-001, 13 April 2018.

Galileo and EGNOS define their 'disclaimer of liability' in their service definition documents for the use of various types of signals.[91]

The content of unilateral disclaimers can be extensive, and the author summarises it in three key points: (i) GNSS providers deny any expressed or implied warranties regarding availability, continuity, accuracy, integrity, reliability, and fitness for a particular purpose or meeting the users' requirements; (ii) GNSS providers readdress that no advice or information, whether oral or written, obtained from any institutes create any such warranty; and (iii) GNSS providers directly state that they are not responsible and liable for any damages and losses including, but not limited to direct, indirect, incidental, special or consequential damages, whether under contractual liability, product liability, strict liability, tort liability or otherwise, including intent or negligence, caused by the use and misuse of signals provided by them.[92]

As a typical safety-critical application of GNSS, EGNOS SoL service nevertheless applies a special disclaimer of civil liability for aviation users, including both airspace users equipped with an EGNOS certified receiver, and certified ANSP having signed a valid EGNOS Working Agreement with the European Satellite Services Provider (ESSP SAS).[93] This disclaimer unilaterally provides warranties specially for the aviation users, and disclaims only *indirect*, *special*, or *consequential* damages resulting from the use of, misuse of, or the inability to use the EGNOS SoL signals,[94] which means that the provider (ESSP SAS) assumes responsibility and liability for any *direct* damage caused by the use of EGNOS SoL signals.[95] Furthermore, the term of contractual liability including similar disclaimers and warranties is addressed again in the EGNOS Working Agreement with each ANSP.

[3] The Enforceability of Disclaimers of Civil Liability in the Context of GNSS

Under the concept of GNSS civil liability, while contract law and tort law do jointly recognise causes of exoneration for GNSS civil liability which mainly refer to 'state of the art technology' and 'acts of God' (*see* §2.03[C]),[96] the term disclaimer of civil

It should be noted that the disclaimer of liability for the use of GNSS signals is different from that for the use of information in the technical documents of GNSS signals.

91. GSA, *European GNSS (Galileo) Initial Services - Open Service - Service Definition Document*, issue 1.0, December 2016, at i [GSA OS]; GSA, *European GNSS (Galileo) Initial Services - SAR/Galileo - Service Definition Document*, issue 1.0, December 2016, at i & ii [GSA SAR]; GSA, *EGNOS Open Service (OS): Service Definition Document*, Revision 2.2, 12/02/2015, at 9 [GSA EGNOS]; GSA, *EGNOS Data Access Service (EDAS): Service Definition Document*, Revision 2.1, 19/12/2014, at 10 [GSA EDAS]; GSA, *supra* note 35, at 9.
92. Cabinet Office of Japan, *supra* note 90; GSA OS, *ibid.*, at i; GSA SAR, *ibid.*; GSA EGNOS, *ibid.*; GSA EDAS, *ibid.*
93. GSA, *supra* note 35.
94. *Ibid.*, at 9.
95. This disclaimer disclaims not only civil liability against the ESSP SAS (EGNOS services provider), but also those against the EU (the owner of EGNOS system), and the European GNSS Agency (GSA, EGNOS Programme manager). *Ibid.*
96. For the distinction between the terms 'acts of God' and 'force majeure', please also *see* footnote 207 in Chapter 2.

liability is not included there. The existence of disclaimers of civil liability may be invoked by GNSS providers as one of the key arguments for barring recovery for defective signals, even though not every disclaimer of liability is legally enforceable. At this time, few international uniform rules concerning disclaimers of civil liability have been discovered. Each State has broad discretion in determining whether or not a disclaimer of civil liability should be upheld or voided in the consideration of national public policy.

A unilateral disclaimer of civil liability tries to avoid GNSS providers having any civil liability, some of which may be caused by the negligence of those providers, through a statement written only by one party, without any negotiations with the other party.[97] Nevertheless, without agreement on its terms concerning disclaimers, one can neither unilaterally impose obligations on another person, nor deprive them of their rights of remedy offered by tort law that presents a sense of public power.[98] The damage caused by defective GNSS signals is not only suffered by the users, but also by innocent third parties for whom a unilateral disclaimer of civil liability is legally pointless. Furthermore, there is no guarantee that the users and third parties will see those unilateral disclaimers stated in technical documents and service definition documents. Those documents may be in the interest of receiver manufacturers or value-added service providers, but not in the interests of end users and third parties who are the parties directly suffering damages. Therefore, the author argues that a unilateral disclaimer of civil liability in the context of GNSS does not have the legal effect of binding another person if he or she has not freely negotiated and consented to it.

The author does not however deny the rationality of the argument that national courts may accept civil liability based on contributory negligence of the claimant - that is GNSS users - based on their decision to use a signal with the awareness of the existence of a warning to use open signals at their own risk, in particular for safety-sensitive applications and the lack of express or implied warranty for the quality of GNSS signals.[99]

The enforceability of disclaimers of civil liability is also a rather difficult question under contract law. Although at this time a contractual relationship between GNSS providers and users is not common (*see* §5.02[C][2]), certain GNSS providers do sign contracts with users, for example the ESSP SAS are required to sign an EWA with an ANSP for the use of EGNOS SoL signals (*see* §5.02[C][4]), where the clauses concerning the disclaimer of civil liability are included (*see* §5.02[D][2]).

A disclaimer of civil liability is a part of the contract, and national courts usually examine its enforceability by weighing the balance between public policy and freedom

97. Standler, *supra* note 52.
98. The reason why GNSS providers make a unilateral disclaimer of civil liability is to fulfil a duty of care that warns users to use the signals at their own risk. However, as discussed in §5.02[C][3], what a GNSS provider is obligated to do is the timely notification of abnormal situations departing from its normal statement on performance parameters in the provision of GNSS signals, rather than a general one-off warning.
99. Chatzipanagiotis & Liperi, *supra* note 71.

to contract based on party autonomy, and the fair allocation of risks between parties.[100] Generally speaking, the courts honour a disclaimer of civil liability if the contract is held sacred and, all other things being equal, except for blatant violations of public policy.[101] Nonetheless, the above doctrine was not altogether sacrosanct.[102] In almost all jurisdictions, legislative reforms intend to protect the interests of consumers - here GNSS users - by either placing restrictions on the excessive use of exclusion clauses, or proscribing certain types of disclaimers.[103] These restrictions mainly refer to the following aspects:

(i) a disclaimer of civil liability only applies to implied warranties;[104] in other words, express warranties cannot be disclaimed.[105] If a GNSS provider has expressly and clearly guaranteed a certain level of performance standard of PNT signals through overt words in contract terms or actions, that provider is deprived of the right to disclaim civil liability for damage caused by defective GNSS signals which were below that promised performance standard. The language of the warranty prevails over that of the disclaimer if the two cannot be reconciled.[106]

(ii) a valid disclaimer of civil liability is not allowed to be against public policy. Most jurisdictions exclude the application of disclaimers of civil liability for any personal injury and the property damage as a result of deliberate intent or gross negligence.[107] Therefore, disclaimers of civil liability in the context of GNSS will be null and void if in their disclaimers those GNSS providers refuse to be held responsible and liable for any personal injury and damage to property caused by gross negligence and deliberate intent of GNSS

100. *See*, e.g., *Atkins v. Swimwest Family Fitness Center*, 691 N.W.2d 334 (Wis. 2005); *Metropolitan Life Ins. Co. v. Noble Lowndes Int'l, Inc.*, 643 N.E.2d 504, 507 (N.Y. 1994); *Tunkl v. Regents of University of California*, 383 P.2d 441 (Cal. 1963); *City of Santa Barbara v. Superior Court*, 41 Cal. 4th 747 (Cal., 2007); *Rosencrans v. Dover Images, Ltd.*, 192 Cal. App. 4th 1072, 1083 (Cal. App., 2011); *Jimenez v. 24 Hour Fitness USA, Inc.* 237 Cal.App.4th 546 (Cal. App., 2015).
101. Chin Nyuk-Yin, *Excluding Liability in Contracts* (Butterworths & Co. (Publishers) Ltd., 1985), at 1 & 8.
102. Henaku, *supra* note 81.
103. Harry Duintjer Tebbens, *International Product Liability: A Study of Comparative and International Legal Aspects of Product Liability* (BRILL, 1979), at 20.
104. An implied warranty is an obligation imposed by the law other than an express promise made by the seller, which only provides a basic level of protection; an express warranty is a definite guarantee, either verbally or in writing, that the product will meet a certain level of quality and reliability. Garner, *supra* note 2, at 1725.
105. Section 2-316 of the US Uniform Commercial Code; *Young & Cooper, Inc. v. Vestring*, 521 P.2d 281 (Kan. 1974).
106. William D. Hawkland, *Limitation of Warranty under the Uniform Commercial Code*, 28(11) Haward Law Journal 1965, at 28.
107. For example, Article 53 of the Contract Law of the People's Republic of China; Item 2, Article 276 of the German Civil Code; Article 332 of the Greek Civil Code; Section 2-719(3) of the US Uniform Commercial Code.

providers. For example, in 'navigation warfare'[108] a GNSS provider terminates the transmission of its PNT signals to deny the use by adversaries, but ignores the impact on civilian users.

(iii) the channelling effect of a disclaimer of civil liability only covers parties on a contractual basis,[109] since a contract cannot create rights or obligations of a third party without its consent.[110] The intention to disclaim civil liability for damage caused to the third party is in vain. Furthermore, a disclaimer must be sufficiently clear, broad, and easily identified in a reasonable way, particular for the case of a format contract or standard term. Otherwise, the enforceability of that disclaimer will be greatly reduced.

[4] Proposal for a Model Clause Concerning Disclaimer of Civil Liability in the Provision of GNSS Safety-of-Life Signals

Theoretically speaking, under the unconscionability standard the equality of parties, both individuals and States, is recognised as a basic principle determining the legally binding force of a contract or agreement or its terms, including a disclaimer of civil liability.[111] In the context of GNSS, the few provider oligarchs seem to have much more bargaining power than the users, in which case an absolute disclaimer of civil liability against all types of damage caused to anyone may be imposed in the political game of negotiation for the provision of GNSS signals. This kind of absolute disclaimer blocks the channelling of GNSS civil liability. Nevertheless, the paramount importance of safety must not be compromised. The vital role of GNSS signals would question the appropriateness of an absolute disclaimer of civil liability in safety-related sectors such as civil aviation.[112]

Therefore, the author argues that a standard clause for the disclaimer of civil liability for damage caused by defective safety-of-life signals may be helpful to protect the recourse right of the innocent party and rechannelling of civil liability to the real wrongdoer, and in turn, put pressure on GNSS providers to provide reliable signals.

As discussed in §5.02[D][2], the ESSP SAS applies a special disclaimer for aviation users in the EWA on the use of EGNOS SoL signals, in which the ESSP SAS only disclaims indirect, special, or consequential damages. This disclaimer is not an absolute disclaimer because it is only related to one type of specific liability, i.e., indirect damage. In other words, the ESSP SAS recognises and accepts civil liability for any direct, physical damage caused by GNSS, even though at the same time it excludes a too heavy burden of civil liability for unexpected damage, including, but not limited to 'damages for interruption of business, loss of profits, goodwill or other intangible losses, resulting from the use of, misuse of, or the inability to use the EGNOS SoL

108. *See* NOAA, *U.S. Space-Based Positioning, Navigation, and Timing Policy: Fact Sheet*, December 15, 2004, https://www.gps.gov/policy/docs/2004/, last accessed 12 June 2018.
109. Andrade, *supra* note 38, at 150.
110. *Cf.* Article 34 of the Vienna Convention.
111. Henaku, *supra* note 81.
112. Andrade, *supra* note 38, at 150.

Service'.[113] Through this disclaimer, the ESSP SAS does not seek to eliminate the warranty of EGNOS SoL signals, but it seeks to control its burden of civil liability in a responsible scope which covers only certain types of damage in the event of a proven breach of contract.[114]

Whether in international law or national legislation, the indirect or consequential damage is a complex and confused concept which generates many legal uncertainties (*see* §3.03[C]). In this regard, the author argues that it is acceptable and reasonable that parties are left free to decide whether indirect or consequential damage is excluded in their disclaimers, as long as the direct or physical damage is recognised by GNSS providers. Furthermore, certain types of damage which are out of the control of GNSS providers, such as the damage caused by uncertified equipment or receivers or a force majeure event, are disclaimed.

Considering the integrity and rationality required by a reasonable disclaimer of civil liability, the author would propose the disclaimer of civil liability made by the ESSP SAS for the provision of EGNOS SoL signals as a model clause, even though the scope of application of this special disclaimer should be extended to all in the safety-critical domains, in addition to the aviation sector.[115]

113. GSA, *supra* note 35, at 9.
114. Tracy, *supra* note 83, at 15.
115. The disclaimer of civil liability reads more or less like:

> By using the EGNOS SoL Service, the Aviation Users agree that neither the European Union nor GSA nor ESSP SAS shall be held responsible or liable for any indirect, special or consequential damages, including but not limited to, damages for interruption of business, loss of profits, goodwill or other intangible losses, resulting from the use of, misuse of, or the inability to use the EGNOS SoL Service.
>
> Furthermore, no party shall be entitled to any claim against ESSP SAS and/or the European Union and/or the GSA if the damage is the result, or the consequence, of any of the following events:
>
> - Use of EGNOS SoL Service beyond the conditions and limitations of use set forth in the EGNOS SoL SDD, it being understood that the use of EGNOS SoL by users other than Aviation Users constitutes a use beyond such conditions and limitations, or
> - Use of equipment or receivers which are
>   - not fully compliant to MOPS (Minimum Operational Performance Standards for Global Positioning System/Wide Area Augmentation System Airborne Equipment) or
>   - not certified or approved by the relevant competent authority or - malfunctioning, or
> - Use of the EGNOS SoL Service when a test message is broadcast (a Message Type 0 or a Message Type 0/2), or
> - Use of the EGNOS SoL Service without required authorisation, or
> - In case of a Force Majeure event.

As the disclosure of EWA documents is restricted for public use, therefore in this section the words for a disclaimer of civil liability contained in those EWA documents are taken from GSA, *supra* note 35, at 9.

[E] Brief Conclusion

There is no such thing as a truly free ride. Not all GNSS signals are against payment. Nonetheless, a free-of-charge policy does not release GNSS providers from civil liability. A policy on charges or a cost-sharing mechanism may force GNSS providers to assume a higher degree of duty of care for applications in safety-critical sectors, and it also eliminates the resistance and unaffordability of GNSS providers to civil liability.

An implied contract does not arise in the provision of open signals, and thus GNSS providers bear no implied contractual liability to users of open signals. In tort law, in the abnormal situation in the provision of open signals GNSS providers merely owe a duty of care to notify users in a reasonable manner, rather than to keep those signals at a safe level. Once GNSS providers perform that duty of care reasonably, GNSS users are no longer entitled to claim any indemnification for further losses. Therefore, the author proposes that an authorisation mechanism on GNSS safety-of-life signals should be established for a legally binding service guarantee containing a stable performance standard meeting the requirement of users such as civil aviation.

A unilateral disclaimer of GNSS civil liability is void since an individual may not be deprived of the right of remedy offered by tort law, which presents a sense of public power. A disclaimer of GNSS civil liability may be upheld, but it may not disclaim: (i) express warranties, (ii) any personal injury and the damage to property as a result of deliberate intent or gross negligence, and (iii) the damage suffered by a third party without its consent. Since at present there is an imbalance in the bargaining power between GNSS providers and users, the author proposes a standard clause for the disclaimer of civil liability for damage caused by defective safety-of-life signals which may be involved as one of the mandatory elements in a contract for the provision of GNSS safety-of-life signals (*see* §5.03[C][2]). For this purpose, the disclaimer of civil liability made by the ESSP SAS for the provision of EGNOS SoL signals merits consideration as a model.

§5.03 THE ROADMAP TO ACHIEVING AN INTERNATIONAL SOLUTION FOR GNSS CIVIL LIABILITY

[A] The Need for an International Solution for GNSS Civil Liability

The fairness of GNSS civil liability has been explained in §5.02. Due to the global nature of GNSS (*see* §1.02[E]) and the international characteristics of GNSS civil liability (*see* §2.05), the international community, in particular the user group, expects a truly international uniform regime for GNSS civil liability, which regime addresses a range of problems such as the doctrine of sovereign immunity, a legal basis of GNSS civil liability, and sustainable development of GNSS industry.[116]

Yet, international air and space law neither offers an adequate solution to ensure sufficient compensation for the victims suffering damage caused by GNSS, nor

116. *Cf.* Ninno, *supra* note 37.

allocates civil liability fairly among stakeholders in the value chain of GNSS (*see* Chapters 3 and 4). Therefore, this section aims to provide a feasible roadmap to achieving an international solution for GNSS civil liability.

[B] Unblocking Sovereign Immunity in the Context of GNSS Civil Liability

[1] The Relevance of Sovereign Immunity

There has been a strong dissenting voice from GNSS providers, typically the US government, on the proposal for an international solution for GNSS civil liability.[117] The author however asserts that domestic rules, in particular those on jurisdictions, are not fully adequate to bring all parties to the court with a view to ensuring fair, prompt, and adequate compensation.[118] One of the key points here is the doctrine of sovereign immunity, which may generally block the right of due process and access to legal remedy. Therefore, in this section the author intends to seek for certain approaches from both legal and institutional perspectives to unblock the doctrine of sovereign immunity in the context of GNSS civil liability.

[2] Sovereign Immunity in the Context of GNSS

Par in parem imperium non habet. On the basis that under international law all States are independent, sovereign, and equal, a sovereign State is disallowed to exercise jurisdiction over another sovereign State before a national court without the latter's approval.[119] This principle has customarily been termed as the doctrine of sovereign immunity.[120] Notwithstanding that international law governs the requirements of this doctrine, its precise extent and manner of application are determined by the individual national law of the State before whose courts a claim against another State is made.[121]

117. *See* Andrade, *supra* note 38, at 87& 88; Kotaite, *supra* note 20.
118. *See also* ICAO, *supra* note 85, at 2.
119. Sompong Sucharitkul, *Immunity of States,* in Mohammed Bedjaoui, International Law: Achievements and Prospects (Martinus Nijhoff Publishers, 1991), at 327.
120. In this book, the author would like to make a distinction between the terms of *foreign sovereign immunity* and *government immunity*. The author argues that the term *foreign sovereign immunity* is based on the maxim *par in parem imperium non habet* (an equal has no power over an equal), and it is a legal doctrine in international law based on *external* sovereignty, which describes the interstate relationship; the term *government immunity* is sourced from the maxim *rex non potest peccare* (the King can do no wrong), and it is a legal doctrine in national law based on *internal* sovereignty, which mainly refers to the immunity from being sued in its own courts without its consent. In international law, *foreign sovereign immunity* is usually worded briefly as the doctrine of sovereign immunity, and this book adopts this practice; however, in the US law, the term sovereign immunity usually refers to government immunity discussed here, which is distinguished from the term foreign sovereign immunity. *See* §1.04[C]; Garner, *supra* note 2, at 818; The US, *Foreign Sovereign Immunities Act of 1976*, October 21, 1976, 90 STAT. 2891, Public Law 94-583, 94th Congress; Erwin Chemerinsky, *Against Sovereign Immunity*, 53 (5) Stanford Law Review 2001, at 1201.
121. Hazel Fox & Philippa Webb, *The Law of State Immunity* (Oxford University Press, 2013), at 1.

In the context of GNSS, the doctrine of sovereign immunity seems to stand out much more than in other space sectors such as radio telecommunication and satellite remote sensing. Most GNSS providers are either military, or civilian authorities of government (*see* §2.04[B]). Therefore, in most cases concerning GNSS civil liability, claimants have to first overcome foreign sovereign immunity at an international level,[122] which prevents national courts at all levels from establishing legal jurisdiction over acts and omissions of another State.[123] Absent specific provisions to the contrary, any claim for civil liability against a State as a responsible GNSS provider in a court outside that State would be impossible and inadmissible,[124] particularly for non-contractual liability.[125] Furthermore, there may also be an obstacle for foreign litigants when a suit is brought against a State GNSS provider in that State's courts.[126]

For example, although courts have proven to be extremely generous towards victims in the US,[127] claims for GPS civil liability against the US government or its Coast Guard, which is the operator of GPS, may easily fail, and the US government has thus avoided many suits based on GPS errors.[128] The US has waived immunity for certain conditions under the Federal Tort Claims Act,[129] but this does not apply to claims arising in a foreign country.[130] In other words, non-US citizens are not allowed to sue the US government in a court outside the US, and they have no choice but to file their claims in a federal court of the US, a situation which may be unaffordable from a practical and political point of view.[131]

[3] Legal Solution: Proposal for a Waiver of Sovereign Immunity in a Legal Instrument

Although each jurisdiction is entitled to determine its regulations on sovereign immunity, citizens suffering damage do deserve an elaborate network of protection.[132] Certain jurisdictions, such as the US and the UK, do not allow a foreign State invoking immunity for personal injuries and damage to property in their courts,[133] but it is still not a general State practice.[134] Further, as discussed, a State usually does not accept the

122. Chatzipanagiotis & Liperi, *supra* note 71, at 171.
123. *Ibid.*; Kantasuk, *supra* note 5, at 48.
124. ICAO, *supra* note 85, at 2; Ninno, *supra* note 37.
125. Baumann, *supra* note 36, at 39; von der Dunk, *supra* note 6, at 141.
126. *Cf.* Alain E. du Perron, *Liability of Air Traffic Control Agencies and Airports Operators in Civil Law Jurisdictions*, 10(4/5) Air & Space Law 1985, at 209.
127. Francis P. Schubert, *Warsaw Claims and ATC Liability: Addressing the Global Dimension of Aviation Liability*, XXXII(I) Annals of Air and Space Law 1997, at 244.
128. Jonathan M. Epstein, *Global Positioning System (GPS): Defining the Legal Issues of Its Expanding Civil Use*, 61 Journal of Air Law and Commerce 1995, at 262-268.
129. 28 U.S.C. Part VI Chapter 171.
130. 28 U.S.C. 2680.
131. von der Dunk, *supra* note 6, at 142.
132. John Mark MacKeigan, *Liability of Air Traffic Services Providers: The Impact of New Systems and Commercialization* (McGill LLM Thesis, 1996), at 42.
133. *See* 28 U.S.C. 1605 (a) (5); Section 5 of the UK State Immunity Act 1978
134. International Court of Justice, *Jurisdictional Immunities of the State (Germany v. Italy: Greece Intervening)*, Judgment of 3 February 2012, I.C.J. Reports 2012, at 27.

claim against it in a foreign court. If the cases concerning GNSS civil liability have to be settled outside of court,[135] it would again bog down claims for compensation in political wrangling and inefficiency (*see* §3.04).

Nevertheless, the fortunate thing is that sovereign immunity may be legally waived through the State's consent to the suit or arbitration.[136] Such consent can either be evidenced by an express declaration or inferred from a State's acts as if immunity had already been waived.[137] The express declaration is usually made before a dispute arises by a written agreement, while the inference can only be made after the dispute has arisen by judging whether the State has involved the claim or defence procedure of any suit - except for the purpose of claiming immunity - voluntarily.[138] Obviously, the express declaration offers better legal certainty for claimants since a State may - and it also has the right to - refuse to intervene in the proceedings, or simply remain silent when facing claims against it in a foreign court.[139]

Therefore, the author proposes that a give-up clause on sovereign immunity could be included if any legal instrument is adopted or concluded for the use of GNSS signals provided by governmental civilian or military authorities. Like the conspicuous requirement for a disclaimer of civil liability (*see* §5.02[D][3]), the said clause must convey "a clear, complete, unambiguous, and unmistakable manifestation of the sovereign's intent to waive its immunity."[140] Such a clause could be either written in a bi/multilateral agreement/commercial contract, or worded in an international convention.

Considering the obstacles for the establishment of a treaty for ATC civil liability (*see* §4.04[D][2]), the author would not expect too much arising from an international convention which also contains a provision on a waiver of sovereign immunity for GNSS civil liability (*see* §5.03[C][1]). Most often, a waiver of sovereign immunity does however arise where there is a contract or an agreement which contains a direct reference to that waiver or a choice of law or an arbitration clause.[141] Furthermore, similar to the situation of ATC service provision (*see* §4.04[D][4]), the commercialisation and corporatisation of GNSS will make such a contract or agreement a bit more feasible (*see* below), even though the State's consent to be sued for the failure of GNSS service is still extremely unlikely in practice.

135. Schubert, *supra* note 127, at 242; Kotaite, *supra* note 20.
136. Ingrid Lagarrigue, *ATC Liability and the Perspectives of the Global GNSS: Is an International Convention Viable?* (McGill LLM Thesis, 1994), at 9.
137. Xiaodong Yang, *State Immunity in International Law* (Cambridge University Press, 2012), at 316.
138. *Ibid.*; Section 2, the UK State Immunity Act 1978; 28 U.S.C. 1605; M. Chase Waring, *Waiver of Sovereign Immunity*, 6 Harvard International Law Club Journal 1965, at 194.
139. *See*, e.g., *Jackson v. People's Republic of China*, 550 F. Supp. 869 (N.D. Ala. 1982).
140. *Aquamar v. Del Monte Fresh Produce*, 179 F.3d 1279, 1292 (11th Cir. 1999).
141. *See*, e.g., *Marlowe v. Argentine Naval Commission*, 604 F.2d 703 (D.D.C. 1985); *Sokaogon Gaming Enter. Corp., et al. v. Tushie-Montgomery Assoc., Inc*, 86 F.3d 656 (7th Cir. 1996); Michael Stoffregen, *Inferred Explicit Standard - Waiver of Sovereign Immunity Via an Arbitration Clause*, 1 Journal of Dispute Resolution 1997, at 155.

[4] *Institutional Solution: Proposal for the Commercialisation and Corporatisation of GNSS*

It is critical to understand that most limits of any meaningful lawmaking efforts on GNSS civil liability arise from the 'natural monopoly' of the current providers and the essentially military roots and nature of most GNSSs.[142] The majority of GNSSs have been national assets first serving the national interests of the provider States.[143] The international community of civilian users can have access only to what the provider State is offering under conditions determined by that State.[144]

To remove those institutional limits, the civil aviation community has worked hard towards achieving a civilian and internationally controlled GNSS over which user States exercise a sufficient level of control on aspects related to its use by civil aviation.[145] Indeed, the industry was vitally interested in the proposal, and the demand from airlines and user States was strong.[146] This notwithstanding, different from the remote sensing sector where the US-Soviet duopoly was rapidly put to an end in the 1980s,[147] no alternative internationally controlled GNSS has been established so far, and this is not foreseen in the near future.[148] The main reason lies in its unaffordability,[149] particularly considering the lack of a policy on charges for GNSS signals.

Therefore, and as further explained below, the author proposes an institutional reform on the commercialisation of GNSS, then movement forward toward the corporatisation of GNSS providers which, more importantly here, will at the same time help unblock the doctrine of sovereign immunity.[150]

(i) The commercialisation of GNSS.

In the 200 years of development,[151] many jurisdictions have moved from the absolute theory of sovereignty immunity toward a restrictive one which excludes the application

142. Milde, *supra* note 3, at 197.
143. von der Dunk, *supra* note 6, at 159.
144. *Cf.* Milde, *supra* note 3, at 212.
145. ICAO, *Global Air Navigation Plan for CNS/ATM Systems*, Second Eidtion-2002, Doc 9750, AN/963, at I-2-8; Milde, *supra* note 3, at 205 & 206; David Sagar, *INMARSAT and GNSS*, 3(1) Outer Space Committee Newsletter 2000, at 41; Jiefang Huang, *Development of the Long-term Legal Framework for the Global Navigation Satellite System*, XXII-I Annals of Air and Space Law 1997, at 596; Ludwig Weber, *Legal and Institutional Issues with Regard to GNSS*, presented to A Conference to examine Legal and Policy Interests Involved in the Implementation of GNSS, ESTEC, Noordwijk, The Netherlands, 14 and 15 November 1996.
146. *Cf.* Milde, *supra* note 3, at 212.
147. Fabio Tronchetti, *Legal Aspects of Satellite Remote Sensing*, in Frans von der Dunk with Fabio Tronchetti, Handbook of Space Law (Edward Elgar Publishing, 2015), at 508.
148. Milde, *supra* note 3, at 212.
149. *Ibid.*, at 205 & 206
150. The proposal for the commercialisation and corporatisation of GNSS applies to both core systems and augmented systems. In a fault liability system, both core systems and augmented systems may be the source of GNSS civil liability.
151. Jasper Finke, *Sovereign Immunity: Rule, Comity or Something Else?*, 21(4) The European Journal of International Law 2011, at 853.

of immunity for commercial activities engaged-in by a State or its government.[152] Regardless of the fact that the debate still exists, the restrictive approach is also gaining currency in international law.[153] Against this background, the commercialisation of GNSS may help to erode the rigid protectionist concept of sovereignty to some degree.[154] However, the existence of a commercial transaction between GNSS providers and users is questioned, especially concerning the provision of free-of-charge open signals which are not contracted for (*see* §5.02[B][2]).[155] Therefore, the commercialisation of GNSS can only be achieved with a policy on charges in an authorisation mechanism (*see* §5.02[B][2] and §5.02[C][4]).

GNSS providers may offer various types of services, for example, commercial service and safety-of-life service. A commercial service mostly constitutes a commercial transaction under the restrictive theory of sovereignty immunity, but the question that arises here is: does a policy on charges really transfer the nature of providing safety-of-life signals from a public service to a commercial service? One analogy is the ANS. It seems that most States still view ANS as a public service even though the ANS charges fees. It is therefore highly questionable and it remains highly uncertain in different jurisdictions whether the provision of GNSS signals, particularly for safety-critical applications, is recognised as a commercial transaction.

In this regard, the author recommends that a contract or an agreement like the EWA should be signed instead of issuing a licence for the use of signals. The reasons include two aspects: (i) a contract is much more connected with private law whereas a licence is usually issued by a sovereign power; and (ii) in certain jurisdictions such as the UK:

> A State is not immune as respects proceedings relating to an obligation of the State which by virtue of a contract (whether a commercial transaction or not) [...].[156]

(ii) The corporatisation of the GNSS provider.

A commercial transaction of GNSS may constitute an exception to sovereign immunity, but a non-sovereign entity on the provision of GNSS signals can generally be sued in any place where they have a presence, since the doctrine of sovereign immunity does not apply to them.[157] It is advisable that such a non-sovereign entity is structured as a corporation with an independent legal personality being capable of assuming civil

152. *See*, e.g., 28 U.S.C. 1605 (a) (2); Section 3(1)(a) of the UK State Immunity Act 1978; Article 5 of the State Immunity Act, Canada. Ernest K. Bankas, *The State Immunity Controversy in International Law: Private Suits Against Sovereign States in Domestic Courts* (Springer Science & Business Media, 2005), at 72; Micheal Brandon, *Sovereign Immunity of Government-Owned Corporations and Ships*, 39(3) Cornell Law Review 1954, at 426.
153. A generic consensus has been reflected in international law by Article 10 of the United Nations Convention on Jurisdictional Immunities of States and Their Property, according to which "the State cannot invoke immunity from that jurisdiction in a proceeding arising out of that commercial transaction," "if the parties to the commercial transaction have expressly agreed otherwise".
154. *Cf.* Milde, *supra* note 3, at 198.
155. Chatzipanagiotis & Liperi, *supra* note 71, at 171.
156. Section 3(1)(b) of the UK State Immunity Act 1978.
157. ICAO, *supra* note 85, at 2.

liability itself. To escape the scope of sovereignty, the corporation would be a 'private', not just a 'civilian' legal entity under private law, which is different from a civilian agency of government.

The corporatisation of the GNSS provider would hopefully help promote international acceptance for users that feel more comfortable dealing with a State enterprise (*see* below), rather than a governmental agency, in particular a military force, which is usually under the protection of sovereign immunity and is subject to political games.[158]

It is nevertheless understandable that certain user groups may distrust the safety and quality of safety-of-life signals provided by a private body other than a governmental agency. Moreover, GNSS provider States may also have concerns about their national security if a GNSS provider is independent of a governmental agency. Such distrust and concerns do exist in the history of GNSS. For example, the augmentation of GPS was partly provided by private suppliers through decryption keys for correction signals for which authorised users had to pay fees.[159] After learning that the US was planning to build a governmental augmentation system of GPS, that is the Nationwide DGPS (NDGPS), these private suppliers argued that the US government should purchase DGPS services from them instead, and they also expressed great resistance since a governmental system would be unfairly competing against them.[160] The US government, however, insisted that none of those private providers are able to meet the technical and reliability requirements imposed by safety-of-life applications.[161] Notably, the US government also pointed that, while it is possible to promulgate regulations that would allow use of commercial services for safety-of-life purposes, there are uncertainties as to whether private providers are willing to accept liability inherent with use of navigation systems.[162]

As a side note, since the majority of governmental augmentation systems based on GPS provide correction signals free-of-charge,[163] it seems almost that all private providers have been priced out of the market; in other words, the lack of a policy on

158. *Cf.* Pace & Frost, *supra* note 7, at 25.
159. *Ibid.*, at 26.
160. *Ibid.*
161. The US Department of Commerce, *A Technical Report to the Secretary of Transportation on a National Approach to Augmented GPS Services*, NTIA Special Publication 94-30, December 1994, at 61.
162. *Ibid.*; Pace & Frost, *supra* note 7, at 26.
163. For example, the NDGPS, the Continuously Operating Reference Station (CORS) developed by the NOAA, the Wide Area Augmentation System (WAAS) developed by the FAA, the DGPS service provided by the Maritime and Port Authority of Singapore (MPA), and the DGPS service provided by the UK Trinity House, the EGNOS Open Service and Safety of Life Service are currently provided free of direct user charge.

See respectively: Jeffrey Auerbach, *GPS Augmentations and Applications*, presented to the Workshop on the Applications of Global Navigation Satellite Systems, Dubai, United Arab Emirates, 16-20 January 2011, at 27; Michael Shaw, et al., *United States Global Positioning System (GPS) and Augmentation Systems Update*, presented to the Ad Hoc Provider's Forum of the International Committee on GNSS, Bangalore, India, 4 September 2007, at 6; NovAtel Inc., *An Introduction to GNSS: GPS, GLONASS, BeiDou, Galileo and Other Global Navigation Satellite Systems* (NovAtel Inc., 2015), at 54; The MPA of Singapore, *Differential Global Positioning System*, https://www.mpa.gov.sg/web/portal/home/port-of-singapore/services/charts-tidal-info-atons-and-hydrography/aids-to-navigation/differential-global-positioning-system, last accessed 25 June 2018; Graham Collins, *Talking about GPS*, https://www.effective-solutions.co

charges will destroy a commercial industry of GNSS, which is usually more cost-effective and innovative than public industry.

In response to those concerns, the author argues that the corporatisation of GNSS should be structured with the following considerations (*see* Figure 5.1):

(i) The corporatisation of GNSS applies only to the provider of GNSS civilian signals. The corporation of GNSS will work only as an interface with civilian users worldwide on the provision of GNSS signals, and it will provide services by a contract under private law instead of a licence issued by a public power (*see also* §5.03[B][4]). Such a corporation will undertake responsibility for concluding contracts, distributing the authorised key, charging user fees if necessary, and assuming civil liability for damage caused by the provision of GNSS signals.

(ii) To ensure national security and safety of signals, which is of paramount importance to each government, the development, construction, operation, and maintenance of GNSS will be the responsibility of, and be funded by the government as part of its national key infrastructure. In addition, national military departments will hold the power to determine the provision of military signals.

(iii) The corporation of GNSS will have a strong link with government, but will not be protected by the doctrine of sovereign immunity. Such corporation could be either a State-owned or a State-controlled enterprise; in other words, States will avail themselves of independent entities under private law to fulfil their tasks.[164] In the provision of safety-of-life signals, the corporation of GNSS will apply the non-profit principle, and all the financial income will be used to improve the safety of signals. For the profit gained from commercial signals, it will make up for the cost of GNSS operations and maintenance.

(iv) The corporation of GNSS will be compulsorily insured for any civil liability in the provision of GNSS signals to maintain sustainable development of GNSS and ensure adequate compensation to victims suffering damage caused by GNSS, since the corporation will usually be a company with limited liability. Alternatively, the government which established that corporation will ensure that a sufficient compensation fund is in place for paying any compensation which is beyond the affordability of the GNSS provider. Otherwise, the State will be responsible and liable for that.

.uk/dgps1.html, last accessed 25 June 2018; GSA EGNOS, *supra* note 91, at 26; GSA, *supra* note 35, at 6.

164. Stephen Hobe, Katharina Irmen & Christian Plingen, *Privatization of German and Other European Air Navigation Service Providers and the Single European Sky Regulations*, XXXII(3) Air & Space Law 2007, at 169.

Figure 5.1 The Proposed Structure of GNSS Corporatisation

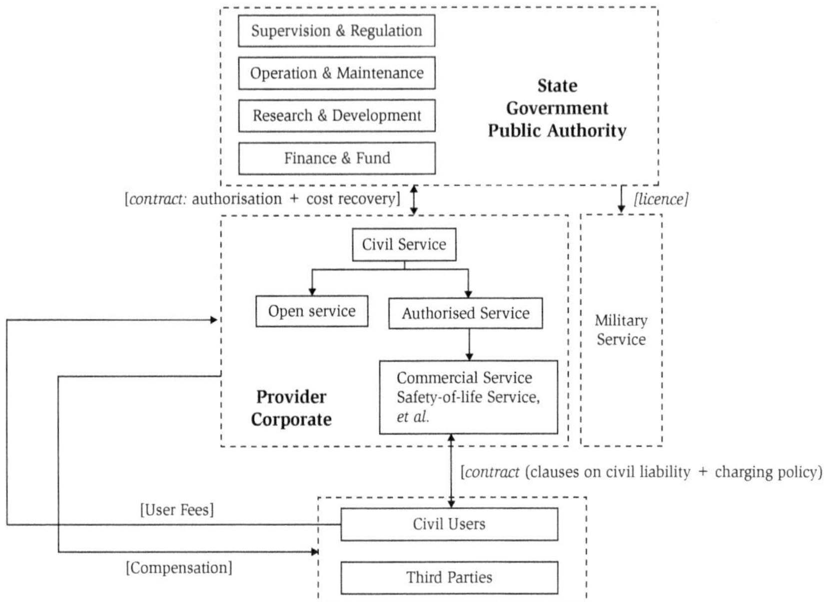

The proposal for the corporatisation of GNSS is being considered under a general tendency both at domestic and international levels towards a separation between a supervisory function – including supervision and certification – and a service provider function. Outside of the context of GNSS, in addition to the reform in the domain of ATC service (*see* §4.04[D][4]), the international space telecommunication community launched a commercialisation and corporatisation action at the eve of this century (*see* Table 5.1); within the context of GNSS, certain GNSS providers have already completed the process of corporatisation:

- the ESSP SAS, as the current service provider of EGNOS and future service provider of Galileo, is "a young and dynamic company specialised in the operations and service provision of safety-critical navigation satellite systems;"[165]
- a group led by the NEC Corporation – a Japanese multinational provider of information technology services and products – was nominated as the service provider of the QZSS, then the group later formed Quasi-Zenith Satellite System Services Inc. (QSS) as the contractor;[166]

165. ESSP, *The ESSP in Brief*, https://www.essp-sas.eu/about-us/#the-essp-in-brief, last accessed in 25 June 2018.
166. Kozuka, *supra* note 89, at 310; Cabinet Office of Japan, *supra* note 90.

- as the co-founder and service provider of GAGAN, Airports Authority of India (AAI) is a corporation rather than a governmental agency, even though it is called 'authority'.[167]

Nevertheless, some service providers of GNSS, such as the QZSS, have another role as the operator, and the author would not support this point due to the consideration of quality and safety of GNSS signals (*see* (ii), above).

Table 5.1 The Corporatisation of International Satellite Communication

Year	*Before the Corporatisation*	*After the Corporatisation*	
		Regulatory Body	*Service Provider*
1999	INMARSAT (International Maritime Satellite Organization)	IMSO (International Mobile Satellite Organization)	Inmarsat Ltd
2001	Eutelsat (European Telecommunications Satellite Organization)	EUTELSAT IGO	Eutelsat S.A.
2001	Intelsat (International Telecommunications Satellite Organization)	ITSO (International Telecommunications Satellite Organization)	Intelsat, S.A.

The author believes that the commercialisation and corporatisation of GNSS will benefit the GNSS industry. The commercialisation and corporatisation of GNSS will promote constructive competition among GNSS providers; more importantly, it will decrease the resistance to GNSS civil liability. For example, the EU is much more positive on GNSS contractual civil liability for civil aviation, and it has adopted a corporatised mechanism for the provision of EGNOS safety-of-life signals (*see* §5.02[D][2]).

[C] The Feasible Approach to GNSS Civil Liability

[1] Various Solutions for GNSS Civil Liability: From a Practical Perspective

GNSS contractual liability, GNSS general tort liability, and GNSS product liability constitute three pillars of GNSS civil liability regime (*see* §2.03[C]). Based on these three pillars, scholars and practitioners have proposed the following approaches to achieving an international solution for GNSS civil liability, even though not all of them are feasible at the current stage.

167. Item (2), Article 3 of the Airports Authority of India Act, No. 55 of 1994, as Amended by the Airports Authority of India (Amendment) Act 2003.

(i) *GNSS general tort liability.* The international community seemingly favours a general tort liability system for GNSS malfunction, with a particular focus on the third-party liability. For example, UNIDROIT has examined the possibility of preparing an international convention for GNSS third-party liability since 2010;[168] the EC published a 'Roadmap on a Regulation on EU GNSS third-party liability' in 2011;[169] the ESPI concluded that the issue of third-party liability should be ruled on as soon as possible with an EU Regulation on GNSS in 2011, proceeding at the same time for an international convention;[170] and European scholars proposed a draft 'Regulation on civil liability and compensation for damage resulting from the activities of European GNSS Services'.[171]

An international instrument of GNSS regulating tort liability, including third-party liability, could be achieved by either an international convention or national legislation, and in the case of Galileo and EGNOS it is an EU regulation. It is rather difficult to urge a large number of States to include GNSS civil liability in their national legislation. Furthermore, the proposed EU regulation has never been published. An international convention, either a specific convention for GNSS civil liability, or a general convention containing articles on GNSS civil liability, is desired by user States,[172] but it has a long way to go, before being achieved, due to strong opposition from provider States,[173] particularly referring to the fate of a convention on ATC civil liability (*see* §4.04[D][2]).

(ii) *GNSS contractual liability.* A contractual framework for GNSS civil liability is continuously promoted by the international civil aviation community as an interim solution towards a final convention.[174] ICAO recognises two models to implement the concept of a contractual framework for GNSS civil liability.[175] Model I is based on a set of non-mandatory contractual clauses in the form of a 'Draft Contractual Framework Relating to the Provision of GNSS Services' (*see* Annex I).[176] Model II is based on a set of mandatory common elements in the form of a 'Framework Agreement between the Governments of [...] Concerning the Implementation, Provision, Operation and Use of a

168. UNIDROIT, *Study LXXIX – Third-Party Liability for Global Navigation Satellite System (GNSS) Services (2010 –)*, https://www.unidroit.org/studies/civil-liability/393-study-lxxix-third-party-liability-for-global-navigation-satellite-system-gnss-services, last accessed 5 July 2018; UNIDROIT, *Item No. 8 on the Agenda: Third-Party Liability for Global Navigation Satellite System (GNSS) Services* (UNIDROIT, 2013), C.D. (92) 8, March 2013.
169. EC, *Roadmap: Regulation on EU GNSS Third-Party Liability*, ENTR.GP2, February 2011.
170. ESPI, *Policy Aspects of Third-Party Liability in Satellite Navigation: Preparing a Roadmap for Europe*, Report 19, P42-C20490-04, July 2009.
171. Sergio M. Carbone, Pietro Manzini, Anna Masutti & Walter Vasselli, *Proposed Regulation on Civil Liability and Compensation for Damage Resulting from the Activities of European GNSS Services*, ALLEGATO 1.
172. ICAO, *supra* note 11, at A-8 to A-10.
173. *Ibid.*
174. ICAO, *supra* note 11, at A-6 to A-8.
175. *Ibid.*
176. ICAO, *supra* note 11, at Attachment F.

Global Navigation Satellite System for Air Navigation Purposes' (*see* Annex II).[177] Each model has its own merits and disadvantages, and the author proposes these two models being implemented in parallel with further developments (*see* §5.03[C][2]).

(iii) *GNSS product liability.* The possibility of classifying a GNSS signal as a 'product' has been examined by analogy with 'electricity' by a few academic researchers,[178] rather than by rule-making bodies (*see* §2.03[C]). The benefits of applying the regime of product liability sourced from the *strict* liability system for damage caused by a product are that this system protects victims in a better way than a fault liability system. Nevertheless, whether courts would recognise an intangible signal as a product presents great legal uncertainties and difficulties,[179] which remains to be resolved by either written law, or case law.

In brief, proposals based on either GNSS general tort liability or GNSS product liability relies too much on the establishment of an international convention, which is very difficult to move further at the current stage. Therefore, the author addresses a more feasible approach - GNSS contractual liability - in the following section.

[2] Proposal for a Contractual Chain on GNSS Civil Liability in the Provision of Safety-of-Life Signals

Different from GNSS tort liability, a contract liability system could be established by the design of the content of a contract between stakeholders in the provision of GNSS signals. The process of adopting an international convention and enacting national legislation is quite a long process with a high risk of failure, but a contract is agreed much more flexibly in shorter-term negotiations. It is however also unrealistic to require GNSS providers to conclude contracts with various groups of users worldwide, particularly considering the public users of open signals (*see* §5.02[C]).[180] Therefore, the author proposes that the regime of GNSS contractual liability focus on safety-of-life service as well as commercial service where there was a clearly worded contract like the EGNOS Working Agreement.

177. ICAO, *Development of a Contractual Framework Leading Towards a Long-Term Legal Framework to Govern the Implementation of GNSS*, A35-WP/125, LE/11, 21/9/04, presented by the 41 Contracting States, Members of the European Civil Aviation Conference, at Appendix B; *Ibid.*, at Attachment G.

178. *See* Andreas Loukakis, *Product Liability Ramifications for Erroneous GNSS Signals: An Alternative Approach Is Possible?*, 56 Proceedings of the International Institute of Space Law 2013, at 320-324; Frans von der Dunk, *GNSS Applications-Legal Implications*, presented to UN Office for Outer Space Affairs, 06-10-2010, at 61-67; Qi Gao, *Civil Liability of GNSS Service Provider: From the Perspective of American Law and Practice*, 29(2) Journal of Beijing University of Aeronautics and Astronautics (Social Science Edition) 2016, in Chinese, at 30-31.

179. *Ibid.*

180. Michael Milde, *Institutional and Legal Problems of the Global Navigation Satellite System (GNSS) - Solutions in Search of a Problem*, in Chia-Jui Cheng & Doo Hwan Kim (Eds), The Utilization of the World's Air Space and Free Outer Space in the 21st Century (Kluwer Law International, 2000), at 352.

As to the contractual chain of GNSS civil liability, there is much to learn from both models recognised by ICAO as discussed above. The key issue of Model I is that it is an optional, non-binding model contract, yet this approach lacks an enforcement mechanism. If there is no other mechanism there to urge GNSS providers to use that model contract, those GNSS providers may ignore that model contract considering their oligopolistic status. Furthermore, although Model I is principally designed for the relationship between the value-added service provider – ANSP and the GNSS signal provider,[181] sometimes the damage caused by GNSS may source from the negligence of GNSS operators, constructors or designers (*see* §2.04[B] and §2.04[C]).

The legal obligation of paying compensation has to be fairly distributed between those stakeholders, and finally channelled to the actual wrongdoer through a set of sequential contracts. Such a series of contracts establishes the legal links for recourse actions which are missing in current international air laws (*see* §4.04[D][5]). Therefore, the author argues that a contractual chain is needed for the allocation of civil liability between all stakeholders,[182] that is each of the two parties in the full liability chain of GNSS development, operation, and service provision (*see* Figure 5.2).[183] Notably, the third-party liability risk may also be allocated between parties by contract, such as the existing contract between the EU and the EGNOS service provider.[184]

181. ICAO, *supra* note 11, at A-7.
182. The discussion here does not include a wrongdoer as a third party which conducts harmful interference (*see* Figure 1-3), in which case no fault may be attributed to GNSS providers.
183. EC, *Impact Assessment on Enhancing the Development and Adoption of Applications of EGNOS and Galileo*, SEC (2010) 717 final, Brussels, 14 June 2010, at 52.
184. David Seïté, *EU GNSS – Third-Party Liability*, presented to International Galileo Governance and Liability Workshop, Transinne, Belgium, 27 May 2011, at 4.

Figure 5.2 Proposed Contractual Chain on GNSS Civil Liability in the Provision of Safety-of-Life Signals

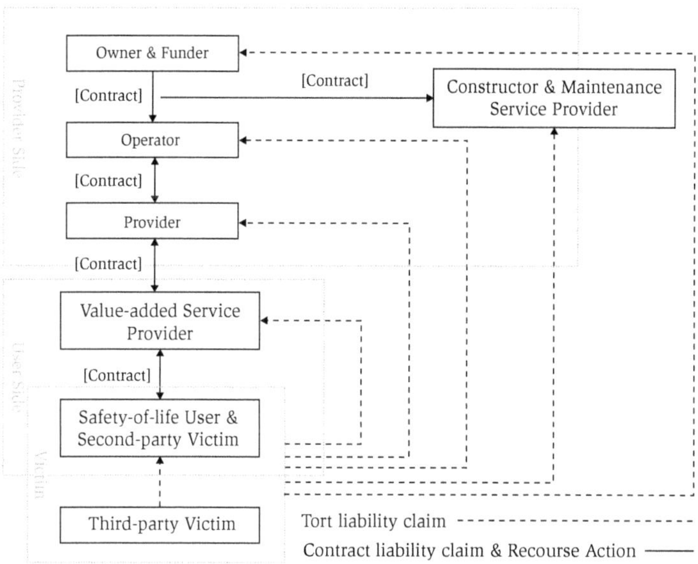

Model II goes beyond Model I to a governmental level.[185] The core of Model II is indeed an intergovernmental agreement signed between a provider State and a user State, which contains some mandatory common elements binding on the parties to achieve the desired degree of uniformity on a series of contracts between private parties in the provision of GNSS signals.[186] This approach makes up for the disadvantageous non-binding nature of Model I.

As there is no constitutional basis which gives ICAO a specific authority to conclude bilateral or multilateral agreements on *external* matters, rather than the internally administrative work of ICAO, with others in the name of the international civil aviation community (*see* §4.03[C]), an intergovernmental agreement has to be agreed among a large number of States, which may require long-term negotiations. Therefore, the author proposes the following mechanism as an interim solution.

For the reasons of safety, IMO and ICAO maintain a sense of control on whether to ratify the standards of a GNSS as parts of their standard system. For example, the BDS has been formally included in the Worldwide Radionavigation System by the IMO's MSC and became the third system to gain such acceptance after the US GPS and Russian GLONASS in 2014.[187] Based on this background, the author suggests that IMO

185. ICAO, *supra* note 11, at A-8.
186. *Ibid.*, at 3.
187. IMO, *SOLAS Amendments to Make IGF Code Mandatory Approved by Maritime Safety Committee, Maritime Safety Committee (MSC)*, 94th session, 17-21 November 2014, http://www.imo.org/en/MediaCentre/PressBriefings/Pages/40-MSC94wrap.aspx#.Wz4Eji-B1lc, last accessed 5 July 2018.

and ICAO may extend that control to recognise 'recommended or designed GNSS providers' for the use of maritime and civil aviation respectively. One of the key conditions for that recognition is that the GNSS provider which submits an application has to comply with the following conditions:

- A GNSS entity (non-sovereign) with an independent legal personality is established under the private law as an interface body to provide safety-of-life signals (*see* §5.03[B][4]).
- A GNSS entity is compulsorily insured for GNSS civil liability, or a sufficient compensation fund is prepared by either this entity or its government (*see* §5.03[B][4]).
- A standard contract is designed for the provision of safety-of-life signals, which should include such mandatory elements as the quality of signals, a policy on charges (*see* §5.02[B][2]), civil liability, and disclaimer of civil liability (*see* §5.02[D][4]).

Due to their military nature, the US GPS and Russian GLONASS, as well as China's BDS, may be reluctant to accept the above proposal, but one solution could be, with the support of the EU's Galileo which may copy the operation structure of EGNOS: establishing a private corporate like the ESSP SAS as the GNSS signal provider; and concluding an agreement like the EGNOS working Agreement between the GNSS provider and users, where civil liability is accepted including a fair disclaimer of civil liability.

[D] The Responsible Organisations to Push the Way Forward

To establish an international uniform governance structure, efforts from three levels are essential: provider level, user level, and supervision level (*see* §1.04[B][3]). The situation is the same in achieving an international solution for GNSS civil liability, where reforms have to be made from the demand, supply, and governance sides.

(i) *From the demand side,* a feasible approach is to make progress first at a chosen unit of GNSS users and then extend the experience learned from that chosen unit to a general scheme. As one of the most sensitive domains for safety, international air transport represents a typical example of reliance on GNSS safety-of-life signals. Indeed, ICAO, as the UN specialised organisation governing international civil aviation, has been working very hard to apply and reform international air law for GNSS technology (*see* §4.02[C]), although no game-changing legal documents have yet been achieved (*see* §4.03). Clear legal gaps in the legal links for making recourse actions possible have however been demonstrated in the present system of international air law when dealing with GNSS civil liability (*see* §4.04 and §4.05). Therefore, ICAO should continue to forge ahead towards an international solution for GNSS civil liability, if possible, by evaluating the roadmap proposed by the author in this chapter. IMO should follow suit in the utilisation of GNSS safety-of-life signals for international maritime transport.

The principal purpose of an international solution for GNSS civil liability is to achieve certain uniformity not only in various jurisdictions but also in different sectors where potential sources of GNSS civil liability are equally varied. It can be expected that driverless vehicles and drones will be other critical domains based on GNSS safety-of-life signals in the near future, and each of them may be subject to different rules. Yet, the sector-specific regime of GNSS civil liability separately developed by ICAO or IMO can only cover a fraction of GNSS applications, and protects a limited group of victims suffering damage caused by GNSS. An international solution for GNSS civil liability should offer a single civil liability regime applicable to all possible fields, also in the case of new applications, and it can only be achieved by the cooperation of all relevant international organisations which, in one way or another, take an interest in GNSS technology and make use of the expertise that each organisation has.[188]

(ii) *From the supply side*, a self-reform of GNSS providers, advocated, hopefully, by the ICG, may eliminate obstacles to an international solution for GNSS civil liability. Without the support and consensus from the supply side, efforts made by the user side will be greatly reduced.[189] For example, the reason why ICAO has not achieved an international instrument with legally binding effect on GNSS civil liability is mainly because of the resistance from primary GNSS providers, in particular, the US government (*see* §4.02[C] and §4.03). An international solution for GNSS civil liability can only be enforceable with the participation of at least one GNSS provider.

The ICG is working as an informal body founded on a voluntary basis, and its object is to promote cooperation in the context of GNSS.[190] Notably, membership of the ICG is not limited to provider States, but also includes international organisations and non-sovereign providers and operators.[191] The tasks of the ICG are organised through four working groups,[192] but legal aspects of GNSS, including the issue of civil liability, have not yet been formally addressed in any of the working groups.[193] To promote the communication and exchanges of information between GNSS providers, the

188. Ninno, *supra* note 37.
189. Milde, *supra* note 3, at 209.
190. ICG, *International Committee on Global Navigation Satellite Systems (ICG)*, http://www.unoosa.org/oosa/en/ourwork/icg/icg.html, last accessed 8 July 2018.
191. ICG, *International Committee on Global Navigation Satellite Systems (ICG): Members*, http://www.unoosa.org/oosa/en/ourwork/icg/members.html, last accessed 8 July 2018.
192. The current groups are (i) Working Group – Systems, Signals and Services; (ii) Working Group – Enhancement of GNSS Performance, New Services and Capabilities; (iii) Working Group – Information Dissemination and Capacity Building; (iv) Working Group – Reference Frames, Timing and Applications. ICG, *International Committee on Global Navigation Satellite Systems (ICG): Working Groups*, http://www.unoosa.org/oosa/en/ourwork/icg/working-groups.html, last accessed 8 July 2018.
193. Baumann, *supra* note 36, at 42.

author proposes that a fifth working group be established, namely, the Working Group – Law and Policy.

Working in parallel with the ICG's Provider's Forum, the Working Group – Law and Policy may contribute based on a voluntary mechanism on the matters of:

- a policy on charges for GNSS safety-of-life signals (*see* §5.02[B][2]);
- an authorisation mechanism of GNSS safety-of-life signals (*see* §5.02[C][4]);
- the commercialisation and corporatisation of GNSS (*see* §5.03[B][4]);
- the model contract and liability chain of the provision of GNSS safety-of-life signals (*see* §5.02[B][2]), including contract terms on a standard disclaimer of civil liability (*see* §5.02[D][4]) and a waiver of State immunity (*see* §5.03[B][3]), if possible.

This notwithstanding, since reform from the supply side as proposed above is based on a voluntary basis, not much achievement can be expected in the near future. An international solution for GNSS civil liability needs the endorsement from the governance side supported by intergovernmental organisations composed of sovereign States.

(iii) *From the governance side,* certain intergovernmental organisations relevant to either GNSS technology, or international law ought to continue moving forward in their efforts to find an international solution for GNSS civil liability. Whether based on an international convention for GNSS tort liability or a governmental framework in an international contractual liability chain, the role played by an international governing body is critical.

GNSS is a space system which applies, among others, international space law (*see* §3.02[A]). At the current stage, the global governance for space affairs is operated under the Committee on the Peaceful Uses of Outer Space (COPUOS), which reports to the General Assembly.[194] The Legal Subcommittee of COPUOS has created a general framework and principal body of international space law, including UN treaties and principles on outer space.[195] These instruments cover comprehensive legal topics ranging from the exploration and use of outer space to space applications such as remote sensing, but no issues related to satellite navigation or GNSS are involved so far.[196] It would be advisable for the Legal Subcommittee of COPUOS to develop legal instruments on GNSS matters, including the issue of civil liability, which could be either treaties, or resolutions. In the author's view, both the proposals for an international convention and a contractual framework discussed in ICAO may also be

194. UNOOSA, *Committee on the Peaceful Uses of Outer Space and its Subcommittees*, http://www.unoosa.org/oosa/en/ourwork/copuos/comm-subcomms.html, last accessed 9 July 2018.

195. Christopher D. Johnson, *United Nations Committee on the Peaceful Uses of Outer Space History, Structure, Agenda, and Current Work*, presented to the International Institute of Air and Space Law, Leiden University, 14 March 2017, at 11; UNOOSA, *Committee on the Peaceful Uses of Outer Space*, http://www.unoosa.org/oosa/en/ourwork/copuos/index.html, last accessed 8 July 2018.

196. UNOOSA, *International Space Law: United Nations Instruments* (United Nations, 2017), at v & vi.

addressed under the framework of the Legal Subcommittee of COPUOS from a general perspective which covers most safety-of-life sectors relying on GNSS.

Furthermore, while serving as the secretariat of COPUOS, since 2005 the United Nations Office for Outer Space Affairs (UNOOSA) has hosted the ICG. The ICG works as an informal body founded on a voluntary basis (*see* above), but it may be naive to expect - although still not impossible to achieve - expansion of the legal competence of the ICG to a regular body or subordinate part of a future specialised agency governing space issues (*see* §1.04[B][3]), especially since the GNSS industry will continue to develop due to the self-reform from the supply side.

Another specialised agency of the UN which is relevant to the operations of GNSS is the ITU. The ITU however mainly contributes to mitigation of harmful interference of GNSS signals through the allocation and coordination of GNSS radio frequencies and satellite orbits.[197] It goes too far to expect the ITU to lead in finding an international solution for GNSS civil liability mainly resulting from the malfunction of the GNSS system.

Outside the UN system, UNIDROIT, an independent intergovernmental organisation, also works on modernising, harmonising, and co-ordinating national private law at an international level through formulating uniform law instruments, principles, and rules.[198] UNIDROIT enjoys a very high reputation for assisting in the development of private international law due to its rich experience in the drafting of international conventions and producing model laws. With reference to GNSS civil liability, in 2010 UNIDROIT started an initial study,[199] after a discussion on the positions for and against preparing an international instrument on the subject.[200] In 2010, UNIDROIT issued a preliminary report on GNSS third-party liability, which concluded that:

> Notwithstanding the existence of a variety of instruments in the transport sector, a number of accidents provoked by GNSS failure or malfunction could fall outside their scope of application.[201]
>
> An international instrument on GNSS liability could provide a sound and uniform regime for all accidents caused by a failure or malfunction of the system, both for those covered and for those not covered by existing international regimes.[202]

After recognising the above legal gap and a need for an international instrument on GNSS civil liability, UNIDROIT planned to move further on this issue. Nonetheless, in 2011 UNIDROIT was informed of the preparation by the EC of an impact assessment intended to evaluate the need for a European regulation on the liability of Galileo. Since

197. ITU, *What Does ITU Do?*, https://www.itu.int/en/about/Pages/whatwedo.aspx, last accessed 8 July 2018.
198. UNIDROIT, *History and Overview*, https://www.unidroit.org/about-unidroit/overview, last accessed 8 July 2018.
199. UNIDROIT, *supra* note 168.
200. *See* respectively: Sergio M. Carbone & Maria Elena De Maestri, *The Rationale for an International Convention on Third Party Liability for Satellite Navigation Signals*, 14 (1-2) Uniform Law Review 2009, at 35-55; Hans-Georg Bollweg, *Initial Considerations Regarding the Feasibility of an International UNIDROIT Instrument to Cover Liability for Damage Caused by Malfunctions in Global (Navigation) Satellite Systems*, 13(4) Uniform Law Review 2008, at 917-934.
201. UNIDROIT, *supra* note 68, at 40
202. *Ibid.*, at 44.

then UNIDROIT has been waiting for that impact assessment before deciding whether to continue its work on GNSS civil liability.[203] Unfortunately, the outcome of that impact assessment has not yet been published.[204] During the waiting period, GNSS industry has made great progress in system development and applications, and more importantly, a multi-GNSS era which is expected to arrive around the decisive year of 2020 (*see* §1.02[B]).

The author argues that now is the time for UNIDROIT to resume the research project of GNSS civil liability. In addition to an international instrument for GNSS third-party liability, UNIDROIT could also address the possibility of a contractual liability chain for GNSS, or a model contract containing such mandatory elements as the quality of signals, policy on charges, civil liability, and disclaimer of civil liability.

International law is drawn up by multiple actors, and no single central organ exercises functions akin to legislatures in States.[205] It is advisable that the Legal Subcommittee of COPUOS and UNIDROIT work jointly as the principal bodies to achieve an international solution for GNSS civil liability. The UN and UNIDROIT can work peacefully together and, more importantly, complement one another.[206] The Legal Subcommittee of COPUOS could coordinate acceptance by GNSS providers States of proposed solutions for GNSS civil liability from a more political mechanism, while UNIDROIT could operate in a more neutral way through an expert-driven approach.

In order to avoid duplication of efforts and encourage more inspiration and support from scholars, both of them should exchange information between each other, and more importantly, work with other academic organisations on international space law, such as the Space Committee of International Law Association (ILA) which issues certain draft conventions on space activities or model laws for national space legislation,[207] the ESPI which showed interested in GNSS third-party liability,[208] and the International Institute of Space Law (IISL) which issues position papers from time to time on certain issues of international space law.[209]

[E] Brief Conclusion

To make an international solution for GNSS civil liability feasible, the first and foremost matter is to find a way to overcome the doctrine of sovereign immunity. Most GNSS providers are now either military, or civilian governmental agencies, in which case any claim for civil liability against a State provider outside that State would be impossible and inadmissible. To avoid this situation, a clause on waiving sovereign immunity may be considered in a legal instrument adopted, or a contract/agreement concluded for the

203. UNIDROIT, *supra* note 168, at 2.
204. Baumann, *supra* note 36, at 44.
205. Miguel de Serpa Soares, *Practicing International Law at the United Nations*, to commemorate the 90th anniversary of UNIDROIT Rome, 15 April 2016, at 7.
206. *Ibid.*, at 4.
207. *See* ILA, *Space Law*, http://www.ila-hq.org/index.php/committees, last accessed 8 July 2018.
208. *See* ESPI, *supra* note 170.
209. *See*, e.g., IISL, *IISL Position Paper on Space Resource Mining*, http://iislweb.org/iisl-position-paper-on-space-resource-mining/, last accessed 9 July 2018.

use of GNSS signals provided by governmental civilian or military authorities. From a long-term perspective, however, the commercialisation of GNSS would make the above approach much more feasible, and more importantly, the cooperation of GNSS providers, not being the operators, on civilian safety-of-life signals would effectively remove the cover of sovereign immunity.

The international community, in particular the international civil aviation community represented by ICAO, has contributed many solutions for GNSS civil liability, even though at the current stage not all of them are feasible. An international instrument for GNSS tort liability, including GNSS product liability, is unlikely due to strong opposition from provider States, particularly referring to the failure practice of a convention on ATC civil liability. Nonetheless, a contractual chain on GNSS civil liability with a focus on the provision of safety-of-life signals seems to be more feasible and flexible through a series of contracts that include provisions on GNSS civil liability, and a governmental framework to ensure some mandatory common elements are included in this contractual chain. Alternatively, ICAO or IMO may take advantage of their power to designate or recommend GNSS suppliers to require these GNSS providers to address GNSS contractual liability in their application.

To push forward towards an international solution for GNSS civil liability, efforts have to be made by relevant organisations from different sides working in parallel:

- *From the demand side*, ICAO and IMO may make headway towards a sector-specific solution, and then extend their experience to a general scheme.
- *From the supply side*, a fifth Working Group on GNSS Law and Policy is proposed to be established through a voluntary mechanism to promote the self-reform of GNSS providers on the commercialisation and corporatisation of GNSS and GNSS contractual liability.
- *From the governance side*, it is advisable for the Legal Subcommittee of COPUOS to develop a resolution on GNSS matters, including the issue of civil liability from a more political mechanism, while UNIDROIT works in a more neutral way through an expert-driven approach.

§5.04 CONCLUDING REMARKS

Although GNSS providers refuse to accept GNSS civil liability due to the problem of open signals with a free-of-charge policy, the fairness of GNSS civil liability to be borne by those providers is not questionable. Nevertheless, to ensure the highest degree of safety for safety-critical applications, an authorisation mechanism for GNSS safety-of-life signals with a policy on charges, where only a restricted disclaimer of civil liability is acceptable, is desirable.

To achieve an international solution for GNSS civil liability, we must currently focus on the provision of safety-of-life signals. With the commercialisation and corporatisation of GNSS, a contractual chain of GNSS civil liability would be an appropriate and feasible approach. GNSS safety-of-life signals determine the public safety of various areas worldwide. An international solution for GNSS civil liability could only be achieved through the cooperation of all relevant international organisations which, in one way or another, take an interest in GNSS technology and make use of each other's expertise.

The lack of legal certainty on GNSS civil liability is more or less delaying the implementation of GNSS. The international community has the freedom on whether or not to introduce GNSS before their concerns were responded to positively. When the oligopoly of GNSS ends and the commercialisation of GNSS begins, the discussion on GNSS civil liability will most likely enter into a new phrase. An international solution for GNSS civil liability, which have been considered unrealistic for decades, will come sooner or later by a gradual convergence from the demand, supply, and governance sides.

ANNEX I Draft Contractual Framework Relating to the Provision of GNSS Services

Attachment F to ICAO A35-WP/75

Whereas the Global Navigation Satellite System (GNSS), as an important element of the communications, navigation and surveillance/air traffic management (CNS/ATM) systems, is intended to provide worldwide coverage and is to be used for aircraft navigation;

Whereas the Parties are desirous to develop the long-term GNSS for civil aviation purposes in accordance with the principles enunciated in the *Charter on the Rights and Obligations of States Relating to GNSS Services*, adopted by the 32nd Session of the Assembly of the International Civil Aviation Organization (ICAO) (A32-19), as set out in the Appendix (hereinafter referred to as the "Charter");

Whereas the Parties aim at ensuring technical and operational accessibility, continuity, availability, integrity, accuracy and reliability of GNSS services;

Whereas the Parties to this contract which are States reaffirm their commitment to act in conformity with international law and the principles governing GNSS, in particular the Convention on International Civil Aviation (Chicago Convention), its Annexes, the Charter and the relevant rules applicable to outer space activities; and the Parties which are not States are committed to act in accordance with applicable law; Therefore, the Parties have agreed as follows:

Article 1 – Parties and Scope of Application

The present contract prescribes the rights and obligations of [*insert Name of Party*], hereinafter "the Air Traffic Service (ATS) Provider", and [*insert Name of other Party*], hereinafter "the GNSS Signal Provider", in respect of all services related to the GNSS signals for the purpose of air navigation. The contract is applicable to the airspace for which the ATS Provider is responsible in relation to its services.

Article 2 – GNSS Signal Provider

For the purposes of the present contract, the term "GNSS Signal Provider" may refer to either:

a) a primary signal provider from the core satellite constellation; or
b) an augmentation signal provider, as the case may be.

Article 3 – Obligations of the GNSS Signal Provider

The GNSS Signal Provider undertakes to:

a) provide the signals during the term of this contract with the required continuity, availability, integrity, accuracy and reliability, as specified in multilaterally agreed standards, in particular the minimum standards of ICAO;
b) if the GNSS Signal Provider is not a State entity, obtain a licence as required by the State in the territory of which the signals are controlled;
c) comply with any requirements arising from the safety management provisions of the relevant Standards and Recommended Practices and Procedures for Air Navigation Services of ICAO; and
d) provide in due time aeronautical information on any modification of the GNSS Signals which may affect the services provided by the ATS provider.

Article 4 – Obligations of the ATS Provider

The ATS Provider undertakes to:

a) if it is not a State entity, obtain from the relevant State the necessary authorisation for the use of GNSS signals provided by the GNSS Signal Provider for air traffic services within the airspace under the jurisdiction of that State;
b) coordinate with the GNSS Signal Provider with a view to facilitating the transmission of the signals and other matters relating to the operation of the GNSS;
c) comply with any requirements arising from the safety management provisions of the relevant Standards and Recommended Practices and Procedures for Air Navigation Services of ICAO; and
d) pay the service charges to the GNSS Signal Provider, if applicable.

Article 5 – Cost Recovery

Pursuant to Article 15 of the Chicago Convention and paragraph 6 of the Charter, the GNSS Signal Provider shall be entitled to establish a cost recovery mechanism, for the purpose of recovering the cost of such services from the users making use of GNSS signals so provided. Such mechanism shall ensure the reasonable allocation of costs among civil aviation users themselves and among civil aviation users and other system users.

Article 6 – Liability

The liability of each Party for failure to perform its obligations under this contract shall be governed by the liability regime applicable to its activity.

Article 7 – Recourse and Indemnification

Nothing in this contract shall prevent any of the Parties from exercising a right of recourse against, or from seeking indemnification from, the other Party or Parties to this contract pursuant to the applicable law.

If the loss or damage has been caused by the acts or omissions of more than one Party, the right of recourse and indemnification of a Party may be limited by the proportion of its respective fault, if the applicable law so provides.

Article 8 – Waiver of Sovereign Immunity

Any Party to this contract which is a State or State entity, hereby agrees to waive its sovereign immunity with respect to any arbitral proceedings in accordance with Article 9 of this contract.

Article 9 – Settlement of Disputes

The Parties shall use their best efforts to settle any dispute, disagreement or claim arising from or relating to the interpretation or performance of this contract by negotiation. Any dispute, disagreement or claim which cannot be settled by negotiation shall be submitted to conciliation in accordance with the UNCITRAL Conciliation Rules.

Any such dispute, disagreement or claim which cannot be settled under the preceding paragraph shall, upon the request of one Party, be referred to arbitration in accordance with the UNCITRAL Arbitration Rules then prevailing. The place of arbitration shall be [...] and it shall be conducted in the [...] language.

Article 10 – Applicable Law

The law of [...] shall govern this contract.

Article 11 – Duration of the Contract

This contract shall enter into force at the date of signature for a term of [...] years and shall be automatically renewable for the same term. Each Party may, however, give notice of termination of [...] months to the other Party, which shall become effective at the end of the term of the Contract.

Article 12 – Registration of the Contract

Pursuant to Article 83 of the Chicago Convention, if at least one Party to this contract is a Contracting State of ICAO, the contract shall be registered with ICAO.

ANNEX II Framework Agreement Between the Governments of Concerning the Implementation, Provision, Operation and Use of a Global Navigation Satellite System for Air Navigation Purposes

Attachment G to ICAO A35-WP/75

1. OBJECTIVES

1.1 The objective of this Agreement is to establish a legal framework for the implementation, provision, operation and use of GNSS for the purpose of air navigation over the territory of Contracting Parties, as well as to regulate the relationships between the entities and persons involved in such GNSS activities.

1.2 This Agreement aims at ensuring technical and operational accessibility, continuity, availability, integrity, accuracy and reliability of GNSS services world-wide. The Contracting Parties reaffirm their commitments to act in conformity with international law and the principles governing GNSS, in particular the Chicago Convention, its Annexes, the Charter on the Rights and Obligations of States Relating to GNSS Services and the relevant rules applicable to outer space activities.

1.3 This Agreement addresses the conditions under which GNSS services, including signals-in-space, can be safely used for air navigation purposes over the territory of Contracting Parties. It also aims at clarifying the obligations of the parties involved.

2. DEFINITIONS

2.1 For the purpose of this Agreement, the terms listed below are used with the following meanings:

Certification: The process which results in a formal attestation that a specified system, element thereof or service complies with pre-determined requirements.

Damage: Loss of life, injury. damage to property [...].

GNSS Entity: A public or private body/organisation, or public-private partnership, created for the purpose of managing, or mandated to manage, by means of contractual arrangements, relationships between GNSS system operators and GNSS service providers involved in the operation of a GNSS system for air navigation purposes.

GNSS service: An added value service to support air navigation, based upon signals emitted by a GNSS system.

GNSS service

provider: An entity engaged in the activity of providing a GNSS service for air navigation purposes.
GNSS signal: A signal emitted by an element forming part of a GNSS system.
GNSS system: An infrastructure comprising satellites and other space and/or ground-based facilities, capable of supporting air navigation based on signals-in-space.

GNSS system

element: Any individual component of a GNSS system.

GNSS system

operator: A body/organisation engaged in the operation and/or maintenance of a GNSS system or elements thereof.
GNSS user: An aircraft which uses GNSS signals or GNSS services for air navigation purposes.

Local augmentation

system: A GNSS system, the purpose of which is to enhance the accuracy, reliability, continuity and integrity of a primary GNSS signal at a given location.

Primary signal

system: A GNSS system, the purpose of which is to produce a primary signal-in-space.

Regional augmentation

system: A GNSS system, the purpose of which is to enhance the accuracy, reliability, continuity and integrity of a primary signal within a given region.

3. SCOPE

3.1 The provisions of this Agreement shall apply to the Contracting Parties implementing, providing, operating and/ or using GNSS for air navigation purposes.

3.2 This Agreement governs the creation of the GNSS Entity or the mandate to an existing entity to perform such function. It addresses, inter alia, the relationships of the Entity with the GNSS system operators and GNSS service providers operating from the territory of a Contracting State or having a registered office on the territory of a Contracting State.

3.3 When Contracting Parties have agreed to undertake responsibilities in respect of providing Air Navigation Services over parts of the high seas, this Agreement shall also apply to the exercise of those responsibilities over those parts of the high seas.

4. SOVEREIGNTY

4.1 This Agreement does not affect in any way the complete and exclusive sovereignty of Contracting Parties in respect of the airspace over their territory.

4.2 The Contracting Parties recognise that the implementation, provision, operation and use of GNSS shall neither infringe nor limit State's authority or responsibility in the control of air navigation and the promulgation and enforcement of safety regulations. States' authority shall also be preserved in the co-ordination and control of communications and in the augmentation, as necessary, of satellite-based Air Navigation Services.

5. CONTRACTING PARTIES RESPONSIBILITIES

5.1 Contracting Parties shall define, in accordance with the provisions of this Agreement, the conditions under which a GNSS system or element thereof may be used for air navigation purposes over their territories.

5.2 Contracting Parties may authorise any public, private or public-private organisations, including foreign bodies, to provide GNSS signals or services to support air navigation over their territory, provided these bodies/organisations operate in accordance with the requirements set forth in this Agreement.

5.3 It remains the responsibility of each Contracting Party to ensure that GNSS signals and services are provided and used over its territory in accordance with the relevant provisions of the Chicago Convention.

5.4 Contracting Parties shall establish appropriate processes:

- to ensure that organisations engaged in the implementation, provision, operation and use of a GNSS system or elements thereof, comply with the requirements of this Agreement; and
- to ensure that the activities performed by the GNSS Entity established or mandated in accordance with article 6 of this Agreement comply with the requirements of this Agreement.

6. GNSS ENTITY

6.1 An Entity shall be established under this Agreement and will be referred to as the GNSS Entity. It shall be made up of an Administrator supported by a Secretariat.

Contracting Parties may mandate an already established organisation or body to undertake the tasks of the GNSS Entity described in this Agreement.

6.2 The GNSS Entity shall have legal personality. It shall enjoy in the territory of its Contracting Parties such legal capacity as may be necessary for the performance of its tasks.

6.3 The GNSS entity shall be charged with facilitating and [managing] [establishing], by means of contractual arrangements, the relationships between the various GNSS system operators and GNSS service providers falling under the scope of this Agreement.

6.4 The GNSS Entity may be entrusted with, inter alia, the following tasks, upon decision by the Contracting Parties:

a) specification of GNSS signals and services;
b) drafting, negotiation, implementation of contractual and service level agreements between the GNSS entity, GNSS system operators and GNSS service providers, in accordance with Article 8 of this Agreement;
c) definition of processes for the allocation of responsibilities among GNSS parties;
d) management of a compensation GNSS fund if set up in accordance with article 9.2 of this Agreement; and
e) definition of applicable risk coverage requirements.

6.5 The financial and institutional consequences of the establishment of the GNSS Entity shall be addressed by the Contracting Parties.

7. ROLE OF ICAO

7.1 Contracting Parties recognise the central role of ICAO in co-ordinating the global implementation of GNSS and in particular:

a) establishment of the SARPs;
b) collection, processing, management and distribution of relevant aeronautical information pertaining to the GNSS systems and services falling within the scope of this Agreement;
c) co-ordination of the activities of the GNSS Entity or body/organisation mandated to undertake its tasks with those of other entities created under similar Agreements and/or with similar functions in other regions; and
d) monitoring of compliance by GNSS system operators and/or service providers with the applicable technical, operational and legal requirements, including the terms of relevant contractual arrangements.

8. CONTRACTUAL AGREEMENTS

8.1 Contracts referred to in Articles 6.3 and 6.4 of this Agreement shall be concluded in conformity with the requirements of this article and the terms of this Agreement.

8.2 Contracting Parties undertake that the contracts entered into in pursuance to this Agreement shall contain the following mandatory elements:

a) compliance with SARPs;
b) compliance with the Charter with regard to continuity, availability, integrity, accuracy and reliability;
c) liability shall be based on fault;
d) compulsory risk coverage;
e) mandatory recourse to arbitration; and
f) recognition that State organisations/bodies are subject to the same rules as private parties.

9. RISK COVERAGE

9.1 The Contracting Parties shall ensure that GNSS system operators and service providers provide adequate insurance or other risk coverage to compensate for loss or damage that may arise out of or in relation to the non-performance of their activities.

9.2 Contracting Parties may set-up a dedicated fund to compensate for any loss or damage that may arise from the non-performance of the activities of system operators or service providers to the extent of a shortfall in the recovery from the body/organisation who is liable.

10. INCIDENT/ACCIDENT INVESTIGATION

10.1 Investigations pertaining to air navigation incidents or accidents involving a possible malfunction, failure or improper use of GNSS shall be conducted in accordance with the provisions of Annex 13 to the Chicago Convention. In this regard, system operators shall ensure that signals shall be recorded for the purposes of evidence.

11. CERTIFICATION

11.1 Contracting Parties shall ensure that GNSS systems and elements including avionics as well as GNSS services shall be certified prior to entry into operation.

11.2 Contracting Parties and their regulators shall ensure, through their established safety management system that GNSS is safe for use. Integrity of the national safety management systems shall be monitored by ICAO [through its Universal Safety Oversight Audit Programme].

12. LIABILITY

12.1 In the event of loss or damage arising out of a failure, malfunction or improper use of GNSS, each entity or person involved shall be liable to the extent it has contributed to the occurrence of the loss or damage.
12.2 The liability of the parties shall be ruled by the material liability regime normally applicable to its activity, in accordance with applicable existing international and national laws.
12.3 Contracting Parties and other public parties shall submit themselves to arbitration and be subject to the same rules as private partners.
12.4 In the event that loss or damage can be attributed to a GNSS failure, malfunction or improper use, but cannot clearly be traced to a specific defendant, the defendants involved in the chain of events which resulted in the occurrence of the loss or damage shall be declared jointly liable for the entire amount of the loss or damage.

13. ARBITRATION

13.1 All liability claims shall be consolidated and brought to arbitration, in accordance with the rules of arbitration established under this Agreement and detailed in Annex [X]. The consolidated claims shall include those against the concerned GNSS Entity, GNSS system operators, GNSS service providers, aircraft operators, air carriers, Air Navigation Services Providers, equipment manufacturers and regulators.
13.2 Nothing in this Agreement shall prejudice the rights of any individuals with regard to the Warsaw/Montreal Conventions.
13.3 Decisions of the arbitration panel shall be final and binding on the Parties to the arbitration procedure.

14. ICAO REGISTRATION

14.1 This Agreement shall be registered with the ICAO Council, in accordance with the provisions of Article 83 of the Chicago Convention.

15. AMENDMENT

15.1 Any proposed amendment to this Agreement shall be subject to the approval of [two-thirds] of its Contracting Parties.

16. ADMISSION OF OTHER PARTIES

16.1 This Agreement is opened for admission to other Parties [...]

17. TERMINATION

17.1 This Agreement may be terminated [...]

Effect on GNSS Entity established under this Agreement [...]

18. ENTRY INTO FORCE

18.1 This Agreement shall enter into force at the date of signature.

General Conclusion

GNSS benefits all walks of life. Major space powers are in various stages of deploying their own GNSSs for the interests of national security and economic growth. Nevertheless, GNSSs are now facing technical, financial, institutional, and legal challenges which may raise various risks on system malfunction or the loss or defection of signals. An integrated governance system consisting of a clear, efficient organisational structure and a comprehensive legal system for GNSS constitutes necessary institutional guarantees and legal safeguards for all innovative solutions to mitigate those risks and related accidents.

Among others, legal certainty on the issue of GNSS civil liability has been an earnest desire of international user community, including both sovereign States and private entities, for decades. But, how does the issue of GNSS civil liability matter? For the answer, the term civil liability has to be understood in the context of GNSS. GNSS is a generic term of core navigation satellite systems and its augmented systems, with global coverage. In the provision of GNSS signals, providers have to follow specific performance parameters, i.e., accuracy, integrity, continuity, and availability, degrading of which damage may be thus caused and the issue of civil liability would arise.

Following a general theory of civil liability, the author defines the term 'GNSS civil liability' as 'the obligation to make reparation for any damage caused, especially in the form of monetary payment, by the inappropriate PNT signal provided by core GNSSs, augmentation systems and regional systems, but excluding the damage caused by GNSS value-added services and the malfunction of user equipment'. GNSS contractual liability, GNSS general tort liability, and GNSS product liability constitute three pillars of the GNSS civil liability regime, which are mostly discussed in legal researches. To generally establish GNSS civil liability, claimants have to prove the existence of four elements, that are the parties, the trigger, the damage, and a causal link.

GNSS civil liability is inherently labelled by its international characteristics in most situations because of transnational litigant parties, cross-border triggers and damage in multiple jurisdictions which are generated by the global coverage and worldwide deployment of GNSS. To deal with such an international civil liability, the domestic approach would cause conflicts of jurisdictions and a series of legal uncertainties. The most critical issue would be the obstacles imposed by the doctrine of sovereignty immunity, particularly because most of GNSS providers are public agencies. It seems that only an international legal framework could better ensure the equitable and uniform compensation for all persons affected irrespective of the State to which they belong.

How does the said international legal framework look like? The international community has been searching for the answer for decades, but, without a consensus. GNSS providers States believe that this international legal framework refers to the present system of international law as there is no indication that it cannot cope with GNSS; while GNSS user States insist that a new international framework has to be developed because no existing international laws may cover the issue of GNSS civil liability. The author concludes the core of this controversy as the question that whether the current system of international law is adequate to deal with the issue of GNSS civil liability? To find out the answer, the applicability of international air and space law to the issue of GNSS civil liability deserves the first try.

International space law regulates the issue of civil liability for damage caused by space objects. Although the development and operation of GNSS qualify as space activities, it is too reluctant to interpret GNSS signals as space objects and hence apply Article VII of the Outer Space Treaty and the Liability Convention. To make civil liability regime under the Outer Space Treaty and the Liability Convention applicable to the issue of GNSS civil liability, some literatures wrongly entangled in the questions of (i) whether indirect damage is covered by Article VII of the Outer Space Treaty and the Liability Convention, and (ii) whether GNSS damage qualifies as indirect damage. The author holds that what really matters lies in the establishment of causation between *damage* and *GNSS satellites*, which qualify the requirement of '*damage* caused by *space objects*'. This causation has to be identified on a case-by-case basis, with the criterion of 'directness', 'foreseeability', and 'proximity' under the sense of general international law.

Nevertheless, the current international space law provides neither an adequate nor a fair mechanism for GNSS civil liability. First, the Liability Convention does not offer any provisions on the juridical proceeding, in the sense of private law, between an individual suffered damage and a liable party, but merely allow that individual to petition his/her government start traditional diplomatic negotiations with the government of the wrongdoer. This State-versus-State liability system involves too many political wrangling. Second, what the current international space law is focusing is civil liability arisen from launching activities, rather than applications of satellites. Third, legal uncertainty may result from different understandings of the matter of causation by different judges and arbitrators, particularly in the case where external factors contribute to the occurrence of damage caused by GNSS.

International air law regulates the issue of civil liability occurred in the course of a flight. Damage caused by the use of GNSS for the purpose of air navigation, as well as automated flight control, falls within the application scope of international air law, ranging from public to private international legal instruments. Because of the inherent connection with the safety of civil aviation, the issue of GNSS civil liability has been a primary focus and concerns of the international civil aviation community.

To establish a legal framework of GNSS, ICAO has released a series of public statements and reports of the Council and its Legal Committee, Assembly Resolutions, SARPs, exchanged letters with States and other documents including manuals, GANPs, and procedures. But, the purpose of those regulatory and guidance materials is mostly for public governance or technical assurance for the implementation of GNSS in the sense of public law; none of those materials addresses legal rules of civil liability directly, even though some of them could be used as reference materials in court to justify the accountability of States providing either GNSS signals or ANS based on GNSS.

GNSS is in essence no more than another navigation aid, but there are many challenges to apply the present regime of ATC civil liability for damage caused by GNSS. First, Article 28 of the Chicago Convention neither enforces any legal obligation on States about the introduction of GNSS for the time being nor serves as the legal basis of ATC or GNSS civil liability under the private law. Second, the national regime of ATC civil liability is reluctant to fit an international civil liability caused by GNSS due to the failure of practice on the proposal for a convention. Third, under the fault-based system of ATC civil liability, victims run the risk of inadequate compensation or no compensation at all in a given case where damage was caused by defective GNSS signals if no negligence can be attributed to the ATC service provider. Fourth, under the no-fault system of ATC civil liability, it is unfair to make an ATC service provider responsible and liable for damage caused by GNSS which is entirely out of its control.

The system of air carrier civil liability is composed of a series of international legal instruments concluded from Warsaw to Montreal via Rome. But none of them provides a legal basis for a direct claim raised by victims against the GNSS provider in the case on GNSS civil liability. If the air carrier is brought in front, a series of recourse actions have to be raised to channel the burden of civil liability to the GNSS provider who is the real wrongdoer. There is no case law giving evidence of the feasibility of such recourse actions. Even in the case ignoring the interest of air carrier but merely focusing on passengers or third parties suffering damage, international air law still cannot ensure those victims receiving full compensation for damage caused by defective GNSS signals in most cases because the majority of conventions regulate a limited civil liability.

In spite of substantial efforts in identifying possible solutions for the issue of GNSS civil liability under international air and space law, neither of them presents much success on the adequacy of victims' compensation and the fairness in the allocation and channel of civil liability. Continuing efforts, coordination and international cooperation are thus needed to search for alternative solutions for the issue of GNSS civil liability. Although the resistance exists from the side of GNSS providers due to the problem of open signals with a free-of-charge policy, and from the protection of the doctrine of sovereignty immunity, the fairness of GNSS civil liability to be borne by those providers is not questionable. Also, the provision of GNSS safety-of-life signals for safety-critical industries deserves much more attention than that of GNSS open signals for the public use in the roadmap to achieve an international solution for GNSS civil liability.

Against this background, the author proposes that efforts have to be made by relevant organisations from different sides working in parallel, particularly in the matters of:

(i) promoting the commercialisation and corporatisation of GNSS through a policy on charges or a cost-sharing mechanism which may force GNSS providers to assume a higher degree of duty of care for safety-critical applications;

(ii) establishing an authorisation mechanism on GNSS safety-of-life signals for a legally binding service guarantee containing a stable performance standard meeting the requirement of users such as civil aviation;

(iii) setting a standard clause for the disclaimer of civil liability for damage caused by defective safety-of-life signals. The disclaimer of civil liability made by the ESSP SAS for the provision of EGNOS SoL signals merits consideration as a model;

(iv) adopting a contractual chain on GNSS civil liability with a focus on the provision of safety-of-life signals through a series of contracts that include provisions on GNSS civil liability, and a governmental framework which contains certain mandatory common elements such as a clause on waiving sovereign immunity.

Bibliography

Books & Thesis

Alessandra A.L. Andrade, *The Global Navigation Satellite System: Navigating into the New Millennium* (Ashgate, 2001).

Alessandra Arrojado Lisboa de Andrade, *Navigating into the New Millennium: The Global Navigation Satellite System Regulatory Framework* (McGill LLM Thesis, 2000).

Andrea J. Harrington, *Regulation of Navigation Satellites in the United States,* in Ram S. Jakhu & Paul Stephen Dempsey (eds), Routledge Handbook of Space Law (Routledge, 2017).

Andreas Loukakis, *EU as Owner of Galileo Satellites: Consequences for Registration and Liability,* in Mahulena Hofmann & Andreas Loukakis (eds), Ownership of Satellites: 4th Luxembourg Workshop on Space and Satellite Communication Law (Nomos, 2017).

Andreas Loukakis, *Non-Contractual Liabilities from Civilian Versions of GNSS: Current Trends, Legal Challenges and Potential* (Nomos, 2017).

Andrew Wilson (Eds), *Galileo: The European Programme for Global Navigation services* (ESA Publication Division, 2005).

Andrzej Górbiel, *Outer Space in International Law* (Uniwersytet Łódzki, 1981).

Anna Masutti, *GNSS: The Basic Principles for a European Legal Framework on TPL,* in Alfredo Roma, Kai-Uwe Schrogl & Matxalen Sanchez Aranzamend (eds), Policy Aspects of Third-Party Liability in Satellite Navigation (ESPI, 2009).

Apostolos Georgiades, *Die Anspruchskonkurrenz im Zivilrecht und Zivilprozeßrecht* (Beck, 1968).

Ariel Porat & Alex Stein, *Tort Liability under Uncertainty* (Oxford University Press, 2001).

Armel Kerrest & Caroline Thro, *Liability for Damage Caused by Space Activities,* in Ram S. Jakhu & Paul Stephen Dempsey (eds), Routledge Handbook of Space Law (Routledge, 2017).

Armmel Kerrest, *Legal Aspects of Transfer of Ownership and Transfer of Activities*, in Mahulena Hofmann & Andreas Loukakis (eds), Ownership of Satellites: 4th Luxembourg Workshop on Space and Satellite Communication Law (Nomos, 2017).

Assuyo Ito, *Legal Aspects of Satellite Remote Sensing* (Koninklijke Brill NV, 2011).

Australian Government Civil Aviation Safety Authority, *CNS/ATM: Transforming Airspace Management* (Civil Aviation Safety Authority, 2017).

Battama Kantasuk, *General Legal Issues Concerning GNSS and the Impact on Developing Countries* (McGill LLM Thesis, 1997).

Ben McFarlane, *The Structure of Property Law* (Bloomsbury Publishing, 2008).

Bernhard Hofmann-Wellenhof, et al., *GNSS – Global Navigation Satellite Systems* (Springer Science & Business Media, 2007).

Bin Cheng, *General Principles of Law as Applied by International Courts and Tribunals* (Cambridge University Press, 1994).

Bin Cheng, *Studies in International Space Law* (Clarendon Press Oxford, 1997).

Boakye Danquah Kofi Henaku, *The Law on Global Air Navigation by Satellite: A Legal Analysis of the ICAO CNS/ATM System* (AST Leiden, 1998).

Bradford W. Parkinson & James J. Spilker et al, *Progress In Astronautics and Aeronautics: Global Positioning System: Theory and Applications-Volume II* (American Institute of Aeronautics and Astronautics, Inc., 1996).

Bran F. Havel & Gabriel S. Sanchez, *The Principles and Practice of International Aviation Law* (Cambridge University Press, 2014).

Brian Harvey, *The Rebirth of the Russian Space Program* (Springer, 2007).

Bruce A. Hurwitz, *State Liability for Outer Space Activities in Accordance with the 1972 Convention on International Liability for Damage Caused by Space Objects* (Martinus Nijhoff Publishers, 1992).

Carl Q. Christol, *Space Law: Past, Present, and Future* (Kluwer Law and Taxation Publishers, 1991).

Carl Q. Christol, *The Modern International Law of Outer Space* (Pergamon Press, 1982).

Carol Rhian Harlow, *Administrative Liability: A Comparative Study of French and English Law* (University of London, 1979).

Cenan Al-Ekabi (Eds), *Yearbook on Space Policy 2012/2013: Space in a Changing World* (Springer-Verlag Wien, 2015).

Chalmers Johnson, *Nemesis: The Last Days of the American Republic* (Metropolitan Books, 2008).

Chia-Jui Cheng (Eds), *Selected Works of Bin Cheng* (Brill Nijhoff, 2018).

Chia-Jui Cheng, *New Sources of International Air Law*, in Chia-Jui Cheng (eds), The Use of Airspace and Outer Space for All Mankind in the 21st Century (Kluwer Law International, 1995).

Chin Nyuk-Yin, *Excluding Liability in Contracts* (Butterworths & Co. (Publishers) Ltd., 1985).

Christian Brünner (Eds), *National Space Law: Development in Europe-Challenges for Small Countries* (Böhlau Verlag Wien, 2008).

Christian von Bar, *Non-Contractual Liability Arising Out of Damage Caused to Another* (sellier. european law publishers, 2009).

Christian von Bar & Ulrich Drobnig, *The Interaction of Contract Law and Tort and Property Law in Europe: A Comparative Study* (sellier. european law publishers, 2004).

Christian von Bar, et al. (Eds), *Principles, Definitions and Model Rules of European Private Law Draft: Common Frame of Reference (DCFR)-Outline Edition* (sellier. european law publishers, 2009).

Christoph Busch, *Secondary Liability for Open Wireless Networks in Germany: Balancing Regulation and Innovation in the Digital Economy*, in Graeme B. Dinwoodie (eds), Secondary Liability of Internet Service Providers (Springer, 2017).

Claire McIvor, *Third-Party Liability in Tort* (Bloomsbury Publishing, 2006).

Clyde Eagleton, *The Responsibility of States in International Law* (Kraus Reprint, 1970).

Committee on Aeronautical and Space Sciences United States Senate, *Convention on International Liability for Damage Caused by Space Objects: Analysis and Background Data-Staff Report* (U.S. Government Printing Office Washington, 1972).

Committee on Aeronautical and Space Sciences United States Senate, *Soviet Space Programs, 1966-70: Goals and Purposes, Organization, Resources, Facilities and Hardware, Manned and Unmanned Flight Programs, Bioastronautics, Civil and Military Applications, Projections of Future Plans, Attitudes Toward International Cooperation and Space Law. Staff Report* (U.S. Government Printing Office, 1971).

Congressional Research Service Library of Congress, *Treaties and Other International Agreements: The Role of the United States Senate* (U.S. Government Printing Office Washington, 2001).

Daniel Rubiano Rincon, *Environmental Law in Colombia* (Kluwer Law International, 2011).

Danny Pieters & Social Security, *An Introduction to the Basic Principles* (Kluwer Law International, 2006).

David M. Walker, *The Oxford Companion to Law* (Clarendon Press, 1980).

Edward G. Lee & David W. Sproule, *Liability for Damage Caused by Space Debris: The Cosmos 954 Claim*, 26 Canadian Yearbook of International Law 1988.

Elie Robert Camille van Bogaert, *Aspects of Space Law* (Kluwer Law and Taxation Publishers, 1986).

Elihu Lauterpacht, Christopher J. Greenwood & Andrew G. Oppenheimer, *International Law Reports (117)* (Cambridge University Press, 2000).

Elliott D. Kaplan & Christopher Hegarty (Eds), *Understanding GPS/GNSS: Principles and Applications* (Artech House, 2017).

Elmar Giemulla & Ronald Schmid (Eds), *Montreal Convention* (Kluwer Law International, 2010).

Elmar Giemulla & Ronald Schmid (Eds), *Montreal Convention: Commentary* (Kluwer Law International, 2010).

Ernest K. Bankas, *The State Immunity Controversy in International Law: Private Suits Against Sovereign States in Domestic Courts* (Springer Science & Business Media, 2005).

Eugene Sochor, *The Politics of International Aviation* (Macmillan Press, 1991).

Ewan McKendrick, *Contract Law: Text, Cases, and Materials* (Oxford University Press, 2014).

FAA, *Advanced Avionics Handbook* (FAA, 2009).

Fabio Tronchetti, *Legal Aspects of Satellite Remote Sensing*, in Frans von der Dunk with Fabio Tronchetti, Handbook of Space Law (Edward Elgar Publishing, 2015).

Francis Lyall & Paul B. Larsen, *Space Law. A Treatise* (Ashgate, 2009).

Francis Schubert, *Air Navigation*, in Paul Stephen Dempsey & Ram S. Jakhu (eds), Routledge Handbook of Public Aviation Law (Routledge, 2017).

Francisco V. Garcia Amador, Louis Bruno Sohn & Richard R. Baxter, *Recent Codification of the Law of State Responsibility for Injuries to Aliens* (Martinus Nijhoff Publishers, 1974).

Frans von der Dunk, *European Space Law*, in Frans von der Dunk with Fabio Tronchetti, Handbook of Space Law (Edward Elgar Publishing Limited, 2015).

Frans von der Dunk, *International Space Law*, in Frans von der Dunk with Fabio Tronchetti, Handbook of Space Law (Edward Elgar Publishing Limited, 2015).

Gabriella Catalano Sgrosso, *International Space Law* (LoGisma editore, 2011).

George N. Tompkins, *Liability Rules Applicable to International Air Transportation as Developed by the Courts in the United States: From Warsaw 1929 to Montreal 1999* (Kluwer Law International, 2010).

Herbert Lionel Adolphus Hart & Tony Honore, *Causation in the Law* (Oxford University Press, 1985).

Harry Duintjer Tebbens, *International Product Liability: A Study of Comparative and International Legal Aspects of Product Liability* (BRILL, 1979).

Helen Gubby, *English Legal Terminology* (Eleven International Publishing, 2016).

Henry G. Schermers & Niels Blokker, *International Institutional Law: Unity Within Diversity* (M. Nijhoff, 1995).

Hermann Mosler, *The International Society as a Legal Community* (Brill, 1980).

Isabella Henrietta Philepina Diederiks-Verschoor & Pablo Mendes de Leon, *An Introduction to Air Law* (Kluwer Law International, 2012).

Isabella Henrietta Philepina Diederiks-Verschoor & Vladimír Kopal, *An Introduction to Space Law* (Kluwer Law International, 2008).

Inge Kaul & Ronald U. Mendoza, *Advancing the Concept of Global Public Goods*, in Pedro Conceição, Katell Le Goulven & Ronald U. Mendoza (eds), Providing Global Public Goods: Managing Globalization (Oxford University Press, 2003).

Ingrid Lagarrigue, *ATC Liability and the Perspectives of the Global GNSS: Is an International Convention Viable?* (McGill LLM Thesis, 1994).

Irmgard Marboe (Ed.), *Soft Law in Outer Space* (Böhlau Verlag Wien Köln Graz, 2012).

Jaap Hage et al. (Eds), *Introduction to Law* (Springer International Publishing Switzerland, 2014).

Jaap Muscle & Francesco Donato Busnelli, *Unification of Tort Law: Liability for Damage Caused by Others* (Kluwer Law International, 2003).

Jean-Pierre Bloch, *La responsabilité des services de la Circulation Aérienne* (Lausanne: Thonney-Dupraz, 1973).

Jeffrey A. Rockwell, *Liability of the United States Arising Out of the Civilian Use of the Global Positioning System* (McGill University, 1996).
Jiefang Huang, *Aviation Safety and ICAO* (Leiden PhD Thesis, 2009).
Jiefang Huang, *Aviation Safety Through the Role of ICAO: ICAO's Mechanisms and Practices* (Kluwer Law International, 2009).
Joan Johnson-Freese, *Space Warfare in the 21st Century* (Routledge, 2017).
Johan Lembke, *Competition for Technological Leadership: EU Policy for High Technology* (Edward Elgar Publishing, 2002).
John Bowen, *Air Transport*, in Jean-Paul Rodrigue, Claude Comtois & Brian Slack, The Geography of Transport Systems (Taylor & Francis, 2016).
John Mark MacKeigan, *Liability of Air Traffic Services Providers: The Impact of New Systems and Commercialization* (McGill LLM Thesis, 1996).
Joseph Chitty & Hugh G. Beale, *Chitty on Contracts* (Sweet & Maxwell, Limited, 2008).
Joseph N. Pelton & Ram Jakhu, *Space Safety Regulations and Standards* (Elsevier, 2010).
Kevin Roebuck, *GLONASS: High-Impact Strategies - What You Need to Know: Definitions, Adoptions, Impact, Benefits, Maturity, Vendors* (Emereo Pty Limited, 2011).
Lawrence B. Goldhirsch, *The Warsaw Convention Annotated: A Legal Handbook* (Kluwer Law International, 2000).
Lesley Jane Smith, *Legal Aspects of Satellite Navigation*, in Frans von der Dunk & Fabio Tronchetti (eds), Handbook of Space Law (Edward Elgar Publishing, 2015).
Ludwig Weber, *The Chicago Convention*, in Paul Stephen Dempsey & Ram S. Jakhu (eds) Routledge Handbook of Public Aviation Law (Routledge, 2017).
Ludwig Weber, *The Global Navigation and Communications Satellite Systems and the Role of ICAO*, in ESA/ECSL, et al., Proceedings of the Third ECSL Colloquium-International Organisations and Space Law (European Space Agency, 1999).
Malcolm Clarke, *Contracts of Carriage by Air* (Lloyd's List, 2010).
Marco Frigessi di Rattalma & Tullio Treves (Eds), *The United Nations Compensation Commission: A Handbook* (Kluwer Law International, 1999).
Marietta Benko, Kai-Uwe Schrogl, Denise Digrell & Esther Jolley, *Space Law: Current Problems and Perspectives for Future Regulation* (Eleven International Publishing, 2005).
Mark E. Villiger, *Commentary on the 1969 Vienna Convention on the Law of Treaties* (Martinus Nijhoff Publishers, 2009).
Marsha L. Baum, *When Nature Strikes: Weather Disasters and the Law* (Greenwood Publishing Group, 2007).
Michael Krauss, *Principles of Products Liability* (West Academic, 2014).
Michael Milde, *International Air Law and ICAO* (Eleven International Publishing, 2012).
Michael Rycroft (Eds), *Satellite Navigation Systems: Policy, Commercial and Technical Interaction* (Springer Science + Business Media Dordrecht, 2003).
Michael S. Nolan, *Fundamentals of Air Traffic Control* (Cengage Learning, 2011).
Michael W. Pearson & Daniel S. Riley, *Foundations of Aviation Law* (Ashgate Publishing, Ltd, 2015).

Nathalie Louisa Johanna Theodora Horbach, *Liability Versus Responsibility under International Law: Defending Strict State Responsibility for Transboundary Damage* (N.L.J.T. Horbach, 1996).
Nandasiri Jasentuliyana & Roy S.K. Lee (Eds), *Manual on Space Law: Volume I* (Oceana Publications, 1979).
Nandasiri Jasentuliyana & Roy S.K. Lee (Eds), *Manual on Space Law: Volume III* (Oceana Publications, 1981).
National Research Council, *Global Navigation Satellite Systems: Report of a Joint Workshop of the National Academy of Engineering and the Chinese Academy of Engineering* (Washington, DC: The National Academies Press, 2012).
Nicolas Mateesco Matte, *Space Activities and Emerging International Law* (McGill University, 1984).
Nicolas Mateesco Matte, *Aerospace Law: From Scientific Exploration to Commercial Utilization* (The Carswell Company Limited & Editions A. Pedone, 1977).
Niels Arnoud van Antwerpen, *Cross-Border Provision of Air Navigation Services with Specific Reference to Europe: Safeguarding Transparent Lines of Responsibility and Liability* (Kluwer Law International, 2007).
Niels van Antwerpen, *Cross-Border Provision of Air Navigation Services with Specific Reference to Europe: Safeguarding Transparent Lines of Responsibility and Liability* (Leiden PhD Thesis, 2007).
NovAtel Inc., *An Introduction to GNSS* (NovAtel Inc., 2010).
NovAtel Inc., *An Introduction to GNSS: GPS, GLONASS, BeiDou, Galileo and Other Global Navigation Satellite Systems* (NovAtel Inc., 2015).
OECD Nuclear Energy Agency, *Liability and Compensation for Nuclear Damage: An International Overview* (OECD Publishing, 1994).
OECD, *Handbook on Measuring the Space Economy* (OECD Publishing, 2012).
OECD, *Space 2030: Tracking Society's Challenges* (OECD Publishing, 2005).
Omri Ben-Shahar & Ariel Porat, *Fault in American Contract Law* (Cambridge University Press, 2010).
Pablo Rodriguez-Contreras Perez, *Damage Caused by GNSS Signals in the Light of the Liability Convention of 1972*, in Michael Rycroft (eds), Satellite Navigation Systems: Policy, Commercial and Technical Interaction (Springer-Science + Business Media, B.V., 2003).
Pablo Mendes de Leon, *100 Years of Air Law - Turning the History of Air Law into Its Future: From Unilateral Acts to International Agreement Through Comity and Diplomacy*, in Stephen Hobe (eds), Air Law, Space Law, Cyber Law – the Institute of Air and Space Law at Age 90 (Carl Heymanns Verlag, 2016).
Pablo Mendes de Leon, *Introduction to Air Law* (Kluwer Law International, 2017).
Pablo Mendes de Leon, *Jurisdiction under and Exclusivity of Private International Air Law Agreements on Air Carrier Liability: The Case of Airbus Versus Armavia Airlines (2013)*, in Pablo Mendes de Leon (eds), From Lowlands to High Skies: A Multilevel Jurisdictional Approach Towards Air Law (Martinus Nijhoff Publishers, 2013).

Pablo Mendes de Leon & Hanneke van Traa, *Space Law*, in Jessica Schechinger (eds), The Practice of Shared Responsibility in International Law (Cambridge University Press, 2017).
Pablo Rodriguez-Contreras Perez, *GNSS Liability Issues: Possible Solutions to a Global System* (McGill University, 2002).
Paul B. Larsen, *Regulation of Global Navigation and Positioning Services in the United States*, in Ram S. Jakhu (ed.), National Regulation of Space Activities (Springer, 2010).
Paul B. Larsen, Joseph Sweeney & John Gillick, *Aviation Law: Cases, Laws and Related Sources* (Martinus Nijhoff Publishers, 2012).
Paul D. Groves, *Principles of GNSS, Inertial, and Multisensor Integrated Navigation Systems* (Artech House, 2013).
Paul Stephen Dempsey, *Aviation Liability Law* (LexisNexis, 2013).
Paul Stephen Dempsey, *Multilateral Conventions and Customary International Law*, in Paul Stephen Dempsey & Ram S. Jakhu (eds) Routledge Handbook of Public Aviation Law (Routledge 2017).
Paul Stephen Dempsey, *Public International Air Law* (McGill University, 2008).
Peter J.G. Tenunissen & Oliver Montenbruck (Eds), *Springer Handbook of Global Navigation Satellite Systems* (Springer International Publishing, 2017).
Peter P.C. Haanappel, *The Law and Policy of Air Space and Outer Space: A Comparative Approach* (Kluwer Law International, 2003).
Peter van Fenema, *The 1972 Outer Space Liability Convention* (McGill University, 1973).
Prashant Popat, *International Product Law Manual* (Kluwer Law International, 2010).
Qizhi He, *Outer Space Law* (Law Press China, 1992), in Chinese.
Aesop, translated by R. Worthington, Aesop's Fables (The Floating Press, 2008).
Ram S. Jakhu et al. (Eds), *The Need for an Integrated Regulatory Regime for Aviation and Space: ICAO for Space?* (Springer-Verlag/Wien, 2011).
Rajeswari Pillai Rajagopalan & Narayan Prasad (Eds), *Space India 2.0: Commerce, Policy, Security and Governance Perspectives* (Observer Research Foundation, 2017).
Richard K. Gardiner, *Treaty Interpretation* (Oxford University Press, 2015).
Robert James Hollyman, *Liability and Reliability: The Reliance Interest in Negligence Damages* (University of Toronto LLM Thesis, 1999).
Ronald I.C. Bartsch, *International Air Law: A Practical Guide* (Ashgate, 2012).
Rupert W. Anderson, *The Cosmic Compendium: Space Law* (Lulu.com, 2015).
Ruwantissa Abeyratne, *Frontiers of Aerospace Law* (Routledge, 2002).
Ruwantissa Abeyratne, *Regulation of Air Transport: The Slumbering Sentinels* (Springer Science & Business Media, 2013).
Ruwantissa Abeyratne, *Rulemaking in Air Transport: A Deconstructive Analysis* (Springer International Publishing Switzerland, 2016).
Ruwantissa Abeyratne, *Space Security Law* (Springer, 2011).
Scott Madry, *Global Navigation Satellite Systems and Their Applications* (Springer, 2015).

Scott Pace & Gerald Frost, et al, *The Global Positioning System: Assessing National Policies* (Rand, 1995).

Sompong Sucharitkul, *Immunity of States*, in Mohammed Bedjaoui, International Law: Achievements and Prospects (Martinus Nijhoff Publishers, 1991).

Stephan Hobe, Bernhard Schmidt-Tedd & Kai-Uwe Schrogl (Eds), *Cologne Commentary on Space Law: Volume II* (Carl Heymanns Verlag, 2013).

Stephan Hobe, Bernhard Schmidt-Tedd & Kai-Uwe Schrogl (Eds), *Cologne Commentary on Space Law: Volume I* (Carl Heymanns Verlag, 2009).

Sushil K. Gupta, *The Routledge Companion to Production and Operations Management* (Taylor & Francis, 2017).

Tare Brisibe, *International Law and Regulation of Aeronautical Public Correspondence by Satellite* (Thesis of Leiden University, 2006).

The International Task Force on Global Public Goods, *Meeting Global Challenges: International Cooperation in the National Interest* (International Task Force on Global Public Goods, 2006).

The State Council Information Office of the People's Republic of China, *China's BeiDou Navigation Satellite System* (Foreign Language Press, 2016).

The Theory Bureau of the Publicity Department of the Central Committee of the Communist Party of China, *Introduction to the Scientific Outlook on Development* (Central Compilation & Translation Press, 2006).

The US Department of State, *Proceedings of the International Civil Aviation Conference: Vol. 1* (United States Government Printing Office, 1948).

Thomas Buergenthal, *Law-Making in the International Civil Aviation Organization* (Syracuse University Press, 1969).

Tom Logsdon, *Understanding the Navstar: GPS, GIS, and IVHS* (Springer, 1995).

UN, *Report of International Arbitral Awards: Mixed Claims Commission* (United States and Germany) (1 November 1923-30 October 1939): Volume VII (United Nations, 2006).

UN, *Reports of International Arbitral Awards* (UN, 1956).

UN, *Yearbook of the International Law Commission 1989: Volume II Part One* (United Nations, 1992).

UN, *Yearbook of the International Law Commission 2000: Volume II Part One* (United Nations, 2009).

UN, *Yearbook of the United Nations 1967* (United Nations, 1969).

United Nations International Law Commission, *Yearbook of the International Law Commission 1998: Documents of the Fiftieth Session (Volume 2 Part 1)* (United Nations Publications, 2009).

UNOOSA, *International Space Law: United Nations Instruments* (United Nations, 2017).

Valerie Kayser, *Launching Space Objects: Issues of Liability and Future Prospects* (Springer Science & Business Media, 2006).

Walter Schwenk & Rüdiger Schwenk, *Aspects of International Co-operation in Air Traffic Management* (Martinus Nijhoff Publishers, 1998).

Wang Hui, *Civil Liability for Marine Oil Pollution Damage: A Comparative and Economic Study of the International, US and the Chinese Compensation Regime* (Kluwer Law International, 2011).
Ward Farnsworth, *Restitution: Civil Liability for Unjust Enrichment* (University of Chicago Press, 2014).
William Benjamin Hale, *Handbook on the Law of Damages* (Рипол Классик, 1896).
William C. Bryson, *Cause and Consequence in the Law*, in Rom Harre & Fathali M. Moghaddam (eds), Questioning Causality: Scientific Explorations of Cause and Consequence Across Social Contexts (ABC-CLIO, 2016).
William Lloyd Prosser, *Prosser and Keeton on the Law of Torts* (West Publishing Company, 1984).
William Lloyd Prosser, *Selected Topics on the Law of Torts: Five Lectures Delivered at the University of Michigan* (William S. Hein, 1982).
World Commission on Environment and Development, *Our Common Future* (WECD, 1987).
Xiaodong Yang, *State Immunity in International Law* (Cambridge University Press, 2012).
Xihui Chen, et al., *A Trust Framework for Evaluating GNSS Signal Integrity*, in Patrick Kellenberger (eds), Proceedings of 2013 IEEE 26th Computer Security Foundations Symposium (The Institute of Electrical and Electronics Engineers, Inc., 2013).
Guangxing Zhang & Shiyuan Han, *General Principles of Contract Law* (Law Press China, 1999), in Chinese.

Dictionaries

Bryan A. Garner (Eds), *Black's Law Dictionary* (WEST, 2009).
Collins COBUILD Advanced Learner's Dictionary (Harper Collins Publishers, 2001).
Elizabeth A. Martin (Eds), *A Dictionary of Law* (Oxford University Press, 2003).
John Bouvier, *Bouvier's Law Dictionary (Vol. II)* (Philadelpia: George W. Childs, 628 & 630 Chestnut St., 1862).
Jonathan Law, *A Dictionary of Law* (Oxford University Press, 2015).
Longman Dictionary of Contemporary English (Foreign Language Teaching and Research Press, 2002).
Macmillan English Dictionary for Advanced Learners of American English (Foreign Language Teaching and Research Press, 2002), at 958.
Sheila Bone (Eds), *Osborn's Concise Law Dictionary* (Sweet & Maxwell Limited, 2001).
The Concise Oxford Dictionary (Foreign Language Teaching and Research Press & Oxford University Press, 1999).
Webster's New World College Dictionary (Liaoning Education Press & Hungry Minds Inc.).

Articles

Aiylam Subramanian Ganeshan, et al., *India's Satellite-Based Augmentation System: GAGAN - Redefining Navigation over the Indian Region*, January/February InsideGNSS 2016.

Alain E. du Perron, *Liability of Air Traffic Control Agencies and Airports Operators in Civil Law Jurisdictions*, 10(4/5) Air & Space Law 1985.

Aldo Armando Cocca, *From Full Compensation to Total Responsibility*, 26 Proceedings of the Twenty-Sixth Colloquium on the Law of Outer Space 1983.

Alexander F. Cohen, *Cosmos 954 and the International Law of Satellite Accidents*, 10(1) Yale Journal of International Law 1984.

Andreas Loukakis, *Product Liability Ramifications for Erroneous GNSS Signals: An Alternative Approach Is Possible?*, 56 Proceedings of the International Institute of Space Law 2013.

Anna Masutti, *Legal Problems Arising from the Installation of the Galileo and EGNOS Ground Stations in Non-EU Countries*, 37 Air & Space Law 2012.

Armel Kerrest, *Remarks on the Responsibility and Liability for Damages Caused by Private Activity in Outer Space*, 40 Proceedings on the Law of Outer Space 1997.

Assad Kotaite, *ICAO's Role with Respect to the Institutional Arrangements and Legal Framework of Global Navigation Satellite System (GNSS) Planning and Implementation*, XXI(II) Annals of Air and Space Law 1996.

Assad Kotaite, *Sovereignty under Great Pressure to Accommodate the Growing Need for Global Cooperation*, 50(10) ICAO Journal 1995.

Bin Cheng, *Centrifugal Tendencies in Air Law*, 10 Current Legal Problems 1957.

Boakye Danquah Kofi Henaku, *The International Liability of the GNSS Space Segment Provider*, XXI(I) Annals of Air and Space Law 1996.

Boakye Danquah Kofi Henaku, *The ICAO CNS/ATM System: New King, New Law?*, XIX(3) Air & Space Law 1994.

Bernard J. Gruber, et al., *Space Superiority, Down to the Nanosecond: Why the Global Positioning System Remains Essential to Modern Warfare*, 5 Air & Space Power Journal 2013.

Bin Cheng, *Article VI of the 1967 Space Treaty Revisited: 'International Responsibility', 'National Activities', and the 'the Appropriate State'*, 26 Journal of Space Law 1998.

Bin Cheng, *International Responsibility and Liability for Launch Activities*, 20(6) Air & Space Law 1995.

Brandon Ehrhart, *A Technological Dream Turned Legal Nightmare: Potential Liability of the United States under the Federal Tort Claims Act for Operating the Global Positioning System*, 33 Vanderbilt Journal of Transnational Law 2000.

Bryan Schwartz & Mark L. Berlin, *After the Fall: An Analysis of Canadian Legal Claims for Damage Caused by Cosmos 954*, 27 McGill Law Journal 1982.

Carl Q. Christol, *International Liability for Damage Caused by Space Objects*, 74 American Journal of International Law 1980.

Caroline Mantl, *Risk Management: The EUROCPNTROL System*, UNIDROIT 2012, C.D. (91) 6, March 2012.

Casey G. Watkins, *Wireless Liability: Liability Concerns for Operators of Unsecured Wireless Networks*, 65(2) Rutgers Law Review 2013.

Chiara Lucchini Gilera, *GNSS Third-Party Liability: The European Experience of Galileo*, 49 Proceeding on the Law of Outer Space 2006.

Chunli Xia, *Study of the ITU Spatial Orbit-Spectrum Resources Allocation and Coordination Rules*, 6 Journal of Beijing Institute of Technology (Social Sciences Edition) 2011, in Chinese.

Cyrille Boulanger, et al., *Receiver Inter System Bias Impact on SBAS Dual Constellation Positioning and Integrity*, Proceedings of the 26th International Technical Meeting of The Satellite Division of the Institute of Navigation (ION GNSS + 2013), Nashville, TN, September 2013.

David Sagar, *INMARSAT and GNSS*, 3(1) Outer Space Committee Newsletter 2000.

Don Tracy, *Disclaiming and Limiting Liability for Commercial Damages*, 83 Commercial Law Journal 1978.

Doo Hwan Kim, *Some Considerations on the Liability of Air Traffic Control Agencies*, 10(1) 한국항공우주정책·법학회지 1988.

Edward P. Warner, *International Air Transport*, 4(2) Foreign Affairs 1926.

Edward R. Finch, *Outer Space Liability: Past, Present and Future*, 14(1) The International Lawyer 1980.

Elena Carpanelli & Brendan Cohen, *Interpreting 'Damage Caused by Space Objects' under the 1972 Liability Convention*, 56 Proceedings of International Institute of Space Law 2013.

Erwin Chemerinsky, *Against Sovereign Immunity*, 53(5) Stanford Law Review 2001.

Evan Criddle, *The Vienna Convention on the Law of Treaties in U.S. Treaty Interpretation*, 44(2) Virginia Journal of International Law 2004.

Federico Bergamasco, *GNSS Liability: Current Legal Framework and Perspectives for the Future from the International Aviation Point of View*, 4 The Aviation & Space Journal 2014.

Frans von der Dunk, *Liability for Global Navigation Satellite Services: A Comparative Analysis of GPS and Galileo*, 30 Journal of Space Law 2004.

Frans von der Dunk, *The International Legal Framework for European Activities on Board the ISS*, in Frans von der Dunk & Marcel M. T. A. Brus, The International Space Station: Commercial Utilisation from a European Legal Perspective (Martinus Nijhoff Publishers, 2006).

Francis P. Schubert, *An International Convention on GNSS Liability: When Does Desirable Become Necessary?*, XXIV Annals of Air and Space Law 1999.

Francis P. Schubert, *Warsaw Claims and ATC Liability: Addressing the Global Dimension of Aviation Liability*, XXXII(I) Annals of Air and Space Law 1997.

Frans von der Dunk, *Liability for Global Navigation Satellite Services: A Comparative Analysis of GPS and Galileo*, 30 Journal of Space Law 2004.

Frans von der Dunk, *Liability Versus Responsibility in Space Law: Misconception or Misconstruction?*, 34 Proceedings of the International Institute of Space Law 1991.

Frans von der Dunk, *Navigating Safely Through the 21st Century: ICAO and the Use of GNSS in Civil Aviation*, 47 India Journal of International Law 2007.

Frans von der Dunk, *Sovereignty Versus Space – Public Law and Private Launch in the Asian Context*, 5 Singapore Journal of International & Comparative Law 2001.
Frans von der Dunk, *Space Law and GNSS – A Look at the Legal Frameworks for 'Outer Space'*, May/June InsideGNSS 2017.
Frans von der Dunk, *Passing the Buck to Rogers: International Liability Issues in Private Spaceflight*, 86 Nebraska Law Review 2007.
Gelli Manoj Someswar, et al., *Global Navigation Satellite Systems and Their Applications*, 17(1) International Journal of Software and Web Sciences 2013.
George M. Cohen, *The Fault that Lies Within Our Contract Law*, 107 Michigan Law Review 2009.
Gregory Michael, *Legal Issues Including Liability Associated with the Acquisition, Use, and Failure of GPS/GNSS*, 52 The Journal of Navigation 1999.
Gregory S. Alexander, *Ownership and Obligations: The Human Flourishing Theory of Property*, 43 Hong Kong Law Journal 2013.
Hamilton DeSaussure & Peter P.C. Haanappel, *A Unified Multinational Approach to the Application of Tort and Contract Principles to Outer Space*, 6 Syracuse Journal of International Law and Commerce 1978.
Hanneke van Traa-Engelman, *Settlement of Space Law Disputes*, 3(3) Leiden Journal of International Law 1990.
Hans-Georg Bollweg, *GNSS-Liability by International or European Union Law?*, 59 German Journal of Air and Space Law 2010.
Hans-Georg Bollweg, *Initial Considerations Regarding the Feasibility of an International UNIDROIT Instrument to Cover Liability for Damage Caused by Malfunctions in Global (Navigation) Satellite Systems*, 13(4) Uniform Law Review 2008.
Helene Sasseville, *Air Traffic Control Agencies: Fault Liability vs. Strict Liability*, X Annals of Air & Space Law 1985.
Hiroyuki Yamada, *IMO and the GNSS: Navigating the Seas*, September/October InsideGNSS 2017.
Ingo Baumann, *Liability for GNSS Signals and Services*, November/December InsideGNSS 2015.
Ingo Baumann, *State of Play in the European Union: Liability for GNSS Signals and Services*, November/December InsideGNSS 2015.
Ingo Baumann, *State of Play in the European Union: Liability for GNSS Signals and Services*, November/December InsideGNSS 2015.
Ingrid Lagarrigue, *Are Existing Navigation Satellite Liability Provisions Adequate to Govern Navigation Satellite Malfunction*, 3(1) Outer Space Committee Newsletter 2000.
Jasper Finke, *Sovereign Immunity: Rule, Comity or Something Else?*, 21(4) The European Journal of International Law 2011.
Jennifer Ann Urban, *Soft Law: The Key to Security in a Globalized Outer Space*, 43 Transport Law Journal 2016.
Jiefang Huang, *Development of the Long-Term Legal Framework for the Global Navigation Satellite System*, XXII(I) Annals of Air and Space Law 1997.
Jingjing Nie, *The Future of Uniform Rules on GNSS Liability*, 54 Proceedings of the International Institute of Space Law 2011.

John A. Lever, *Unintended Consequences of the Global Positioning System*, 7 Systems Engineering 2004.

John M. Kelson, *State Responsibility and the Abnormally Dangerous Activity*, 13 Harvard International Law Journal 1972.

Jonathan M. Epstein, *Global Positioning System (GPS): Defining the Legal Issues of Its Expanding Civil Use*, 61 Journal of Air Law and Commerce 1995.

Jonathan M. Epstein, *Global Positioning System (GPS): Defining the Legal Issues of Its Issues of Its Expanding Civil Use*, 61 Journal of Air Law and Commerce 1995.

Kim Doo Hwan, *Some Considerations on the Liability of Air Traffic Control Agencies*, XIII(6) Air & Space Law 1988.

Kim Murray, *The Law Related to Satellite Navigation and Air Traffic Management Systems - a View from the South Pacific*, 31 VUW Law Review 2013.

Larry W. Thomas, *Liability of State Highway Departments for Design, Construction, and Maintenance Defects*, 4 Selected Studies in Highway Law 1978.

Lazar Vrbaski, *Liability of Air Navigation Services Providers: Towards an International Solution*, 38(1) Air & Space Law 2013.

Ludwig Weber & Jiefang Huang, *ICAO and GNSS*, 3(1) Outer Space Committee Newsletter 2000.

Ludwig Weber, *Convention on International Civil Aviation - 60 Years*, 53(3) German Journal of Air and Space Law 2004.

Luke Punnakanta, *Space Torts: Applying Nuisance and Negligence to Orbital Debris*, 86 Southern California Law Review 2012.

M. Chase Waring, *Waiver of Sovereign Immunity*, 6 Harvard International Law Club Journal 1965.

M.J. Asbury, *Some Institutional Factors and Aspects Relating to a Civil Global Navigation Satellite System*, 47(2) Journal of Navigation 1994.

Marco Ferrazzani, *The Role and Liabilities of Space Segment Operators*, in European Centre of Space Law, Regulation of the Global Navigation Satellite System (GNSS): A Conference to examine Legal and Policy interests involved in the implementation of GNSS (ESTEC, 14-15 November 1996).

Martin Grzebellus, *Is Certification of Galileo a Bureaucratic Overhead?*, 4 Coordinates 2008.

Melvin Aron Eisenberg, *The Role of Fault in Contract Law: Unconscionability, Unexpected Circumstances, Interpretation, Mistake, and Nonperformance*, 107 Michigan Law Review 2009.

Michael Chatzipanagiotis & Konstantina Liperi, *Regulation of Global Navigation Satellite Systems*, in Ram S. Jakhu & Paul Stephen Dempsey (eds), Routledge Handbook of Space Law (Routledge, 2017).

Michael Milde, *Air Navigation and Safety: Institutional and Legal Problems of the Global Navigation Satellite System*, IV Temas de aviacion comercial y derecho aeronautico y espacial 2000.

Michael Milde, *Institutional and Legal Problems of the Global Navigation Satellite System (GNSS) - Solutions in Search of a Problem*, in Chia-Jui Cheng & Doo Hwan Kim (eds), The Utilization of the World's Air Space and Free Outer Space in the 21st Century (Kluwer Law International, 2000).

Michael Milde, *Institutional and Legal Problems of the Global Navigation Satellite System (GNSS) - Solutions in Search of a Problem*, in Chia-Jui Cheng & Doo Hwan Kim (eds), The Utilization of the World's Air Space and Free Outer Space in the 21st Century (Kluwer Law International, 2000).

Michael Milde, *Solutions in Search of a Problem? Legal Problems of the GNSS*, XXII(II) Annals of Air and Space Law 1997.

Michael Stoffregen, *Inferred Explicit Standard - Waiver of Sovereign Immunity Via an Arbitration Clause*, 1 Journal of Dispute Resolution 1997.

Micheal Brandon, *Sovereign Immunity of Government-Owned Corporations and Ships*, 39(3) Cornell Law Review 1954.

Muhamed Mustaque, *Legal Aspects Relating to Satellite Navigation in Air Traffic Management with Specific Reference to GAGAN in India*, 50 Proceedings on the Law of Outer Space 2008.

Nick Ward, *Monitoring the Integrity of GNSS*, 47(2) Journal of Navigation 1994.

Nie Jingjing, *The Future of Uniform International Rules on GNSS Liability*, 54 Proceedings of the International Institute of Space Law 2011.

Norman Bonnor, *A Brief History of Global Navigation Satellite Systems*, 65 Journal of Navigation 2012.

Paul B. Larsen, *Air Traffic Control: A Recommendation for a Proof of Fault System Without a Limitation on Liability*, 32 Journal of Air Law and Commerce 1966.

Paul B. Larsen, *UNIDROIT Space Protocol: Comments on the Relationship Between the Protocol and Existing International Space Law*, 44 Proceedings of the International Institute of Space Law 2001.

Paul Stephen Dempsey, *The Future of International Air Law in the 21st Century*, 64 German Journal of Air and Space Law 2015.

Paula Duggan Vraa & Steven M. Sitek, *Public Policy Considerations for Exculpatory and Indemnification Clauses: Yang v. Voyagaire Houseboats*, 32(4) William Mitchell Law Review 2006.

Peter H. Sand, *The International Unification of Air Law*, 30(2) Law and Contemporary Problems 1965.

Pietro Manzini & Anna Masutti, *An International Civil Liability Regime for the Galileo Services: A Proposal*, 33 Air & Space Law 2008.

Piotr Manikowski, *Examples of Space Damages in the Light of International Space Law*, 6(1) The Poznań University of Economics Review 2006.

Prachee Kulkarni & Pablo Mendes de Leon, *Liability for Damage in Civil Aviation Within the Context of GNSS*, 3(1) Outer Space Committee Newsletter 2000.

Qi Gao, *Civil Liability of GNSS Service Provider: From the Perspective of American Law and Practice*, 29(2) Journal of Beijing University of Aeronautics and Astronautics (Social Science Edition) 2016, in Chinese.

Ra Michael Chatzipanagiotis & Konstantina Liperi, *Regulation of Global Navigation Satellite Systems*, in Ram S. Jakhu & Paul Stephen Dempsey (eds), Routledge Handbook of Space Law (Routledge, 2017).

Ram S. Jakhu & Steven Freeland, *The Relationship Between the United Nations Space Treaties and the Vienna Convention on the Law of Treaties*, 55 Proceedings of the International Institute of Space Law 2012.

Ram S. Jakhu & Steven Freeland, *The Relationship Between the Outer Space Treaty and Customary International Law*, 59 Proceedings of the International Institute of Space Law 2016.

Ram S. Jakhu, Diane Howard & Andrea J. Harrington, *Legal Aspects of Solar Power Satellites*, in Leslie I. Tennen (eds), Private Law, Public Law, Metalaw and Public Policy in Space (Springer International Publishing Switzerland, 2016).

Ranjana Kaul, *Liability Implications of the Use of Global Navigation Satellite Systems (GNSS) for Communication, Navigation, Surveillance/Air Traffic Management (GNS/ATM) in Civil Aviation: With Special Focus on India*, XXXV(I) Annals of Air and Space Law 2000.

Ranjana Kaul, *Liability Implications of the Use of Global Navigation Satellite Systems (GNSS) for Communication, Navigation, Surveillance/Air Traffic Management (GNS/ATM) in Civil Aviation: With Special Focus on India*, XXXV(I) Annals of Air and Space Law 2000.

Richard A. Epstein, *The Many Faces of Fault in Contract Law: Or How to Do Economics Right, Without Really Trying*, 107 Michigan Law Review 2009.

Robert A. Anthony, *Interpretive Rules, Policy Statements, Guidances, Manuals, and the Like - Should Federal Agencies Use Them to Bind the Public?*, 41(6) Duke Law Journal 1992.

Roderick D. van Dam, *GNSS and Aviation: Eurocontrol's Perspective*, Outer Space Committee Newsletter, 2000.

Sang Wook Daniel Han, *Global Administrative Law: Global Governance of the Global Positioning System and Galileo*, 14 ILSA Journal of International & Comparative Law 2008.

Serge Plattard, *Can Global Navigation Satellite Systems Signals Qualify to Become a World Public Good?*, 3(3) New Space 2015.

Sergey Karutin, *System for Differential Correction and Monitoring Updated*, Proceedings of the 24th International Technical Meeting of the Satellite Division of the Institute of Navigation (ION GNSS 2011), Portland, OR, September 2011.

Sergio M. Carbone & Maria Elena De Maestri, *The Rationale for an International Convention on Third Party Liability for Satellite Navigation Signals*, 14(1-2) Uniform Law Review 2009.

Sofia Michaelides-Mateou, *Customary International Law in Aviation: A Hundred Years of Travel through the Competing Norms of Sovereignty and Freedom of Overflight*, in Brian D. Lepard (eds), Reexamining Customary International Law (Cambridge University Press, 2017).

Solène Rowan, *Fault and Breach of Contract in France and England: Some Comparisons*, 22(4) European Business Law Review 2011.

Souichirou Kozuka, *Regulation of Navigational Satellites in Japan*, in Ram S. Jakhu & Paul Stephen Dempsey (eds), *Routledge Handbook of Space Law* (Routledge, 2017).

Statement by Deputy Press Secretary Speakes on the Soviet Attack on a Korean Civilian Airliner, 16 September 1983, in Ronald Reagan, Public Papers of the Presidents of the United States: Ronald Reagan 1983 (United States Government Printing Office, 1984).

Stephan Hobe, Katharina Irmen & Chiristian Plingen: *Privatization of German and Other European Air Navigation Service Providers and the Single European Sky Regulations*, XXXII(3) Air & Space Law 2007.

Stephen Gorove, *Some Thoughts on Liability for the Use of Data Acquired by Earth Resources Satellites*, 15 Proceedings of the International Institute of Space Law 1972.

Stephen Gorove, *Toward a Clarification of the Term 'Space Object': An International Legal and Policy Imperative?*, 21(1) Journal of Space Law 1993.

Stephen Hobe, Katharina Irmen & Christian Plingen, *Privatization of German and Other European Air Navigation Service Providers and the Single European Sky Regulations*, XXXII(3) Air & Space Law 2007.

Steven Reinhold, *Good Faith in International Law*, 2 UCL Journal of Law and Jurisprudence 2013.

Ulf Linderfalk, *Is the Hierarchical Structure of Article 31 and 32 of the Vienna Convention Real or Not? Interpreting the Rules of Interpretation*, 54(1) Netherlands International Law Review 2007.

Ulrich Magnus, *Civil Liability for Satellite-Based Services*, 13 Uniform Law Review 2008.

Victor Iatsouk, *Development of Standards for Aeronautical Satellite Navigation System*, 54 Acta Astronautica 2004.

Vladimir Kopal, *Some Remarks on Issues Relating to Legal Definitions of 'Space Object', 'Space Debris' and 'Astronaut'*, 37 Proceedings of the International Institute of Space Law 1999.

William F. Foster, *The Convention on International Liability for Damage Caused by Space Objects*, 10 The Canadian Yearbook of International Law 1972.

William Michael Reisman, *Soft Law and Law Jobs*, 2(1) Journal of International Dispute Settlement 2011.

William D. Hawkland, *Limitation of Warranty under the Uniform Commercial Code*, 28(11) Haward Law Journal 1965.

Presentations & Lectures

Alessandro del Ninno, *Providing GNSS Services: The Legal Perspective. The Existing Regime and Its Shortcomings with Regards to Liability, Data Policy and Data Integrity*, presented to The International Workshop GNSS Technology Advances in a Multiconstellation Framework, Rome, 26 September 2014.

Anna Masutti, *CSN/ATM Systems: Framework and Regulation on GNSS, Experiences in Europe*, presented to the Conference on Contemporary Issues in Air Transport, Air Law and Regulation, New Delhi, India, 21-25 April 2008.

Christopher D. Johnson, *United Nations Committee on the Peaceful Uses of Outer Space History, Structure, Agenda, and Current Work*, presented to the International Institute of Air and Space Law, Leiden University, 14 March 2017.

David Seïté, *EU GNSS – Third-Party Liability*, presented to International Galileo Governance and Liability Workshop, Transinne, Belgium, 27 May 2011.

Durk van Willigen, *Radio Navigation: Perspectives and Challenges*, presented to NAV02 – GNSS Vulnerability, London, 5-7 November 2002.

Francis Schubert, *The Liability of Air Traffic Control Agencies - The Ueberlingen Midair Collision Case Study*, Presentation to the Institute of Air and Space Law, McGill University, October 2014.

Francis Schubert, *The Liability of Air Traffic Control Agencies – The Ueberlingen Midair Collision Case Study*, presented to the International Institute of Air & Space Law, Leiden University, 11 October 2017.

Frans von der Dunk, *GNSS Applications – Legal Implications*, presented to Training Course on Global Navigation Satellite Systems and Location Based Services, African Regional Centre for Space Science and Technology Education in English (ARCSSTE-E), Ile Ife, Nigeria, 4-29 October 2010.

Gerhard Bertler, et al., *GLONASS and Multi-GNSS in the IGS: Lessons Learned from GLONASS Service Disruptions*, presented to 13th Meeting of the National Space-Based Positioning, Navigation, and Timing (PNT) Advisory Board, Washington DC, 3-4 June 2014.

Gerhard Berz, *Authorisation and Operation of GNSS Aviation Services in Non-Core Constellation States*, presented to Civil GPS Service Interface Committee (CGSIC), Tampa, USA, 8 September 2014.

Jeffrey Auerbach, *GPS Augmentations and Applications*, presented to the Workshop on the Applications of Global Navigation Satellite Systems, Dubai, United Arab Emirates, 16-20 January 2011.

Jim Nagle, *ICAO Policy on GNSS, GNSS SARPs and Global GNSS Developments*, presented to ICG-2, Bangalore, India, 5-7 September 2007.

Ludwig Weber, *Legal and Institutional Issues with Regard to GNSS*, presented to A Conference to examine Legal and Policy Interests Involved in the Implementation of GNSS, ESTEC, Noordwijk, The Netherlands, 14 and 15 November 1996.

Legal Aspects of GNSS, AN-Conf/11-WP/143, 18/9/13, presented by African States to Eleventh Air Navigation Conference, Montreal, 22 September to 3 October 2003.

Michael Shaw, et al., *United States Global Positioning System (GPS) and Augmentation Systems Update*, presented to the Ad Hoc Provider's Forum of the International Committee on GNSS, Bangalore, India, 4 September 2007.

Miguel de Serpa Soares, *Practicing International Law at the United Nations*, lectured to commemorate the 90th anniversary of UNIDROIT Rome, 15 April 2016.

Mohamed Smaoui, *ICAO Global Provisions and Regional Developments Related to GNSS*, presented to Joint ACAC/ICAO MID Workshop on GNSS, Rabat, Morocco, 7-8 November 2017.

Oleg Denisenko, *Proposals on the Development of the International GNSS Monitoring and Assessment System*, presented to Working Group Meeting of ICG, Prague, Czech Republic, 11-13 November 2014.

Ranjana Kaul, *India-Liability in Context to the Air Navigation Service Provider*, presented to the International Conference on Contemporary Issues in Air Transport, Air Law & Regulation, New Delhi, India, 23-25 April 2008.

Urs Hugentobler, et al., *IGS Activities on GLONASS*, presented to GGOS-RAS MEETING, Vienna, 9 April 2013.

Yaw Out Mankata Nyampong, *Privatization and/or Commercialization of Air Navigation Services and Its Liability Implications*, presented to the International Conference on Contemporary Issues in Air Transport, Air Law and Regulation, New Delhi, India, 23-25 April 2008.

Table of Legislation

Legislation & Policy

Treaties & Agreements

Documents & Reports

UN

ICAO

Definition of International Standards and Recommended Practices, A1-31, 1947., 120
Development and Elaboration of an Appropriate Long-Term Legal Framework to Govern the Implementation of GNSS, A32-20, 1998., 104
Development of a Contractual Framework Leading Towards a Long-Term Legal Framework to Govern the Implementation of GNSS, A35-WP/125, LE/11, 21/9/04, presented by the 41 Contracting States, Members of the European Civil Aviation Conference., 105, 185
Directives to Divisional-Type Air Navigation Meetings and Rules of Procedure for Their Conduct, Doc 8143-AN/873/3, 1983., 120
Draft Agreement Between the International Civil Aviation Organization (ICAO) and GNSS Signal Provider Regarding the Provision of Signals for GNSS Services, in Doc. 9630-LC/189., 111
Electronic Bulletin EB 2014/035, 15 July 2014., 147
Evolution of Guidelines for Regional Legal Framework to Govern the Implementation of GNSS (SBAS), A36-WP/134, LE/6, 6/9/07, presented by India., 105
Global Air Navigation Plan for CNS/ATM Systems, Doc 9750, AN/963, Second Edition-2002., 41, 178
Global Air Navigation Plan, Doc 9750, AN/963, Third Edition-2007, 18, 23
Global Coordinated Plan for Transition to ICAO CNS/ATM Systems, Appendix to the Report of the Fourth Meeting of the Special Committee for the Monitoring and Co-ordination of Development and Transition Planning for the Future Air Navigation System (FANS Phase II), Doc 9623, 1993., 101
Global Navigation Satellite System (GNSS) Manual, Doc 9849, AN/457, Second Edition-2013, 9, 34, 99, 157
Legal Aspects of GNSS, AN-Conf/11-WP/143, 18/9/03, presented by the African States., 106
Operational Concept for Next Generation GNSS, PBNICG/4 – IP/06, Agenda Item 11, 15/03/17, presented by Secretariat., 130
Performance-Based Navigation (PBN) Manual, Doc 9613, AN/937, Third Edition-2008., 9
Preliminary Draft – International Convention on the Liability of Air Traffic Control Agencies, submitted by the Republic of Argentina to the 25th Session of the ICAO Legal Committee (Montreal, 12-27 April 1983)., 139
Procedures for Air Navigation Services: ICAO Abbreviations and Codes, Doc 8400, 2010., 122
Progress in the Work of the Panel of Legal and Technical Experts on the Establishment of a Legal Framework with Regard to GNSS (LTEP), A23-WP/24, LE/3, 18/6/98., 104
Progress Report on the Establishment of a Legal Framework with Regard to GNS/ATM Including GNSS, A33-WP/34, LE/5, 22/06/01., 104,
Provisional Agenda - Plenary Meetings, A34-WP/1, P/1, 24/1/03., 104
Rationalization of the ICAO SARPs System, A36-WP/246, TE/80, 18/9/07, presented by the Civil Air Navigation Services Organisation (CANSO)., 138
Regional Multinational Organisms, A36-WP/231, LE/8, 18/9/07, presented by Colombia., 105

Report of the 10th Air Navigation Conference, Doc. 9583, 1991., 100
Report of the 28th Session of the ICAO Legal Committee, Doc 9630-LC189, 1992., 103
Report of the 29th Session of the ICAO Legal Committee, Doc. 9630-LC/189., 112
Report of the Fourth Meeting of the Special Committee on Future Air Navigation Systems, Doc. 9524, FANS/4, Rec. 2/1, 1988., 100
Report of the Legal Commission for the General Section of Its Report and on Agenda Items 7, 8, 45, 46, 47 and 48, A36-WP/341, P/49, 25/9/07., 105
Report of the Rapporteur on the 'Consideration, with Regard to Global Navigation Satellite System (GNSS), of the Establishment of a Legal Framework', LC/29-WP/3-1, 3 March 1994, presented by Kenneth Rattray., 112
Report of the Tenth Air Navigation Conference, Doc. 9583, AN-CONF/10., 111
Report on the Establishing of a Legal Framework with Regard to CNS/ATM Systems Including GNSS, A35-WP/75, LE/5, 28/07/04., 104, 157
Resolution of the Legal and Institutional Aspects Associated with the Global Implementation of GNSS, A35-WP/215, LE/17, 27/09/04, presented by the IATA., 102
Statement of ICAO Policy on CNS/ATM Systems Implementation and Operation, ICAO Doc. WP/3-2, LC/29, 28 March 1994., 26
The Current Status and Further Development of the GLONASS Orbital Grouping in Support of Multi-Constellation GNSS Implementation, A39-WP/452, TE/199, 20/9/16, presented by the Russian Federation., 124
Working Paper C-WP/13478, the 188th session of ICAO Council, 7/10/09., 147
World-wide CNS/ATM Systems Implementation Conference Report, Doc 9719, May 1998., 103

EC & ESA

EC, *Communication from the Commission to the European Parliament and the Council: State of Progress of the Galileo Programme*, COM/2002/0518 final, 2002/C 248/02, 15 October 2002., 34
EC, *Impact Assessment on Enhancing the Development and Adoption of Applications of EGNOS and Galileo*, SEC (2010) 717 final, Brussels, 14 June 2010., 186
EC, *Proposal for a Regulation of the European Parliament and of the Council on Further Implementation of the European Satellite Navigation Programmes (2014-2020)*, SEC (2011) 1446 final, Brussels, 30 November 2011., 51
EC, *Roadmap: Regulation on EU GNSS Third-Party Liability*, ENTR.GP2, February 2011., 39, 184
ESA, *Declaration of Acceptance of the Astronauts Agreement, the Liability Convention and the Registration Convention*, adopted by the ESA Council on 12 December 1978, deposited on 2 January 1979., 89
European GNSS Programmes Committee, *Global Navigation Satellite Systems (GNSS): Extra Contractual Liability*, EGPC-09-07-06-02, 24 June 2009., 39
GSA, *EGNOS Data Access Service (EDAS): Service Definition Document*, Revision 2.1, 19 December 2014., 169
GSA, *EGNOS Open Service (OS): Service Definition Document*, Revision 2.2, 12 February 2015., 169

GSA, *EGNOS Safety of Life (SoL) – Service Definition Document*, Revision 3.1, 26 September 2016., 161

GSA, *European GNSS (Galileo) Initial Services – Open Service – Service Definition Document*, issue 1.0, December 2016., 169

GSA, *Press Release: GSA Earns ISO 9001 Certification*, GSA/PR/14/09, Prague, 11 December 2014., 18

Others

Cabinet Office of Japan, *Quasi-Zenith Satellite System – Interface Specification – Positioning Technology Verification Service*, IS-QZSS-TV-001, 13 April 2018., 168

China Satellite Navigation Office, *BeiDou Navigation Satellite System Signal In Space Interface Control Document Open Service Signal (Version 2.0)*, December 2013., 36

China Satellite Navigation Office, *Report on the Development of BeiDou Navigation Satellite System (Version 2.2)*, December 2013, 93.

Civil Aviation Department of the Government of the Hong Kong Special Administrative Region, *Safety Management Systems (SMS) for Air Operators and Maintenance Organizations: A Guide to Implementation.*, 109

ESPI, *Policy Aspects of Third-Party Liability in Satellite Navigation: Preparing a Roadmap for Europe*, Report 19, P42-C20490-04, July 2009., 46, 184

Eurocontrol, *EUROCONTROL Policy on GNSS for Navigation Applications in the Civil Aviation Domain*, Action Paper, SCG/8/AP10, 28 April 2008., 130

Eurocontrol, *GNSS Sole Service Feasibility Study*, EEC Note No. 04/03, Project GNS-Z-SBAS., 164

Eurocontrol, *PBN Implementation Issues: Adapting SES Framework to GNSS*, Version 5, 6 September 2013., 17

German Federal Bureau of Aircraft Accidents Investigation, *Investigation Report*, Ax001-1-2/02, May 2004., 85, 108, 139

IMO, *Maritime Policy for a Future Global Navigation Satellite System (GNSS)*, RESOLUTION A.860(20), A20/Res.860, 2 December 1997., 21

IMO, *SN.1/Circ/329.*, 21

IMO, *SN/Circ.182.*, 21

IMO, *SN/Circ.187.*, 21

IMO, *Worldwide Radionavigation System*, Resolution A.1046(27), A 27/Res.1046, 20 December 2011., 21

International Court of Justice, *Jurisdictional Immunities of the State (Germany v. Italy: Greece Intervening)*, Judgment of 3 February 2012, I.C.J. Reports 2012., 176

International Court of Justice, *Reparation for Injuries Suffered in the Service of the Nations, Advisory Opinion*, [1949] ICJ Rep 174., 113

Japan Aerospace Exploration Agency, *Quasi-Zenith Satellite System Navigation Service: Interface Specification for QZSS (IS-QZSS)*, IS-QZSS V1.6, 28 November 2014., 7

Karl von Wogau, *On Space and Security*, 2008/2030(INI), Report of 10 June 2008 to European Parliament, A6-0250/2008., 16

Sergio M. Carbone, Pietro Manzini, Anna Masutti & Walter Vasselli, *Proposed Regulation on Civil Liability and Compensation for Damage Resulting from the Activities of European GNSS Services*, ALLEGATO 1., 184

UNCC Governing Council Decision No. 7, S/AC.26/1991/7/Rev.1, 17 March 1992., 81

Unidroit, *An Instrument on Third Party Liability for Global Navigation Satellite System (GNSS) Services: A Preliminary Study*, UNIDROIT 2010, Study LXXIX - Preliminary Study, March 2010., 46, 76, 103, 165

Unidroit, *Item No. 7 on the Agenda: Third-Party Liability for Global Navigation Satellite System (GNSS) Services*, Unidroit 2012, C.D. (91) 6, March 2012., 144, 158

Unidroit, *Item No. 8 on the Agenda: Third Party Liability for Global Navigation Satellite System (GNSS) Services*, UNIDROIT 2013, C.D. (92) 8, March 2013., 184

US Congress, *A BILL: To Require the Secretary of Defense to Establish a Backup for the Global Positioning System, and for Other Purposes*, H.R.1678, 114th Congress (2015-2016)., 6

US Department of Commerce, *A Technical Report to the Secretary of Transportation on a National Approach to Augmented GPS Services*, NTIA Special Publication 94-30, December 1994., 180

US Department of Defense, *Preservation of Continuity for Semi-Codeless GPS Applications*, Federal Register / Vol. 73, No. 185 / Tuesday, 23 September 2008 / Notices., 6

US Government Accountability Office, *Global Positioning System: A Comprehensive Assessment of Potential Options and Related Costs Is Needed*, GAO-13-729., 14

US Government Accountability Office, *GPS: Actions Needed to Address Ground System Development Problems and User Equipment Production Readiness*, Report to Congressional Committees, GAO-15-657, September 2015., 18

Table of Cases

Cases

Index

H

I

AEROSPACE LAW AND POLICY SERIES

1. Erwin von den Steinen, *National Interest and International Aviation* (ISBN 90-411-2455-1).
2. Ludwig Weber, *International Civil Aviation Organisation: An Introduction* (ISBN 978-90-411-2622-1).
3. Niels van Antwerpen, *Cross-Border Provision of Air Navigation Services with Specific Reference to Europe: Safeguarding Transparent Lines of Responsibility and Liability* (ISBN 978-90-411-2688-7).
4. Brian F. Havel, *Beyond Open Skies: A New Regime for International Aviation* (ISBN 978-90-411-2389-3).
5. Jiefang Huang, *Aviation Safety through the Rule of Law: ICAO's Mechanisms and Practices* (ISBN 978-90-411-3115-7).
6. Marian Hoeks, *Multimodal Transport Law: The Law Applicable to the Multimodal Contract for the Carriage of Goods* (ISBN 978-90-411-3246-8).
7. George N. Tompkins, Jr., *Liability Rules Applicable to International Air Transportation as Developed by the Courts in the United States: From Warsaw 1929 to Montreal 1999* (ISBN 978-90-411-2646-7).
8. Daniel Calleja Crespo and Pablo Mendes de Leon (eds), *Achieving the Single European Sky: Goals and Challenges* (ISBN 978-90-411-3730-2).
9. Donal Patrick Hanley, *Aircraft Operating Leasing: A Legal and Practical Analysis in the Context of Public and Private International Air Law*, Second Edition, (ISBN 978-90-411-6050-8).
10. Mikołaj Ratajczyk, *Regional Aviation Safety Organisations: Enhancing Air Transport Safety through Regional Cooperation* (ISBN 978-90-411-5861-1).
11. Benjamyn I. Scott (ed.), *The Law of Unmanned Aircraft Systems: An Introduction to the Current and Future Regulation under National, Regional and International Law* (ISBN 978-90-411-6126-0).
12. Stamatis Varsamos, *Airport Competition Regulation in Europe* (ISBN 978-90-411-6831-3).
13. Katarzyna Malinowska, *Space Insurance: International Legal Aspects* (ISBN 978-90-411-6784-2).
14. John Milligan, *European Union Competition Law in the Airline Industry* (ISBN 978-90-411-6618-0).
15. Dejian Kong, *Civil Liability for Damage Caused by Global Navigation Satellite System* (ISBN 978-94-035-1171-9).

www.ingramcontent.com/pod-product-compliance
Ingram Content Group UK Ltd.
Pitfield, Milton Keynes, MK11 3LW, UK
UKHW020249290926
15470UKWH00002B/826

9 789403 511719